Conference Board of the Mathematical Sciences

CBMS

Regional Conference Series in Mathematics

Number 100

Calderón-Zygmund Capacities and Operators on Nonhomogeneous Spaces

Alexander Volberg

Published for the
Conference Board of the Mathematical Sciences
by the
American Mathematical Society
Providence, Rhode Island
with support from the
National Science Foundation

CBMS Regional Research Conference
Nonhomogeneous Harmonic Analysis, Weights, and
Applications to Problems in Complex Analysis and Operator Theory
University of North Carolina
May 13–17, 2002

Partially supported by the National Science Foundation

2000 *Mathematics Subject Classification.* Primary 42B20;
Secondary 32A55, 31A15, 31C05.

For additional information and updates on this book, visit
www.ams.org/bookpages/cbms-100

Library of Congress Cataloging-in-Publication Data
Volberg, Alexander, 1956–
    Calderón-Zygmund capacities and operators on nonhomogeneous spaces / Alexander Volberg.
    p. cm. — (Regional conference series in mathematics, ISSN 0160-7642 ; no. 100).
    Includes bibliographical references.
    ISBN 0-8218-3252-2 (alk. paper)
    1. Calderón-Zygmund operator. I. Title. II. Series.

QA1 .R33 329   no. 100
[QA329.2]
510 s—dc22   2003062990
[515′.7246]

Contents

CHAPTER 1

Introduction

In 1994-1995 a certain hope was raised on the possibility of describing removable sets of bounded holomorphic functions (the so-called Painlevé problem, 100 years old; or Vitushkin's problem, approximately 50 years old). Also there was a feeling that the problem of Vitushkin concerning the semi-additivity of analytic capacity (approximately 50-year-old problem as well) could be solved. This hope was based on essentially two things: 1) on the impetus given to the research in this area from works of Melnikov [**37**] and Melnikov–Verdera [**39**], and 2) on a certain analytic approach, which was gaining popularity due to the works of Jones [**25**]–[**27**], Coifman–Jones–Semmes [**11**], Christ [**5**], and Murai [**38**] from the late 80's.

This approach (which, in fact, led to the solution of all those problems) was based on using the real analysis methods to these complex analysis questions. Actually, it was based on using $T1, Tb$ theorems from the theory of Calderón–Zygmund operators. However, in the theory of Calderón–Zygmund (CZ) operators, $T1, Tb$ theorems were proved in the so-called homogeneous space setting, when the underlying measure has the doubling condition. Actually, $T1, Tb$ were first proved in the Lebesgue measure setting by David–Journé [**12**], [**13**], David [**15**], [**16**], and David–Journé–Semmes [**14**]. The generalization of $T1, Tb$ theorems from the Lebesgue measure setting even to the homogeneous space setting turned out to be already quite difficult. This generalization was obtained by Christ in [**5**] exactly for the purpose of approaching Vitushkin's conjectures.

This homogeneous setting seemed to be so natural that it was considered one of the cornerstones of the theory. However, the essence of the new approach to the above-mentioned problems required getting rid of this cornerstone entirely. This is just because Vitushkin's conjectures deal with very irregular sets with no homogeneity.

And this "getting rid of" is exactly what happened. Independently, and exactly at the same time, in [**61**] and in [**43**], the nonhomogeneous $T1$ theorem was proved.

Tolsa's proof was tailored to the need of Vitushkin's problems, that is to Cauchy singular integral operator (it used the notion of curvature of measure related to the Cauchy kernel), and Nazarov–Treil–Volberg's proof was for general Calderón–Zygmund operators on nonhomogeneous spaces. This generality turned out to be very essential for the approaches to Vitushkin's conjectures that follow.

However, nonhomogeneous $T1$ was only the first step in the right direction, because Vitushkin's problems require nonhomogeneous Tb theorems rather than $T1$ theorems. These types of results were obtained by David–Mattila [**18**] and David [**19**]. The geometric description of removable sets for bounded analytic functions (the problem of Painlevé, or Vitushkin) followed. But semiadditivity of analytic capacity could not be derived from this result because it had qualitative, not quantitative, nature. This was a very big road block. Also the abovementioned nonhomogeneous Tb theorem had a horrendously tough proof, which made slightly unrealistic the hope to get its quantitative version.

Very soon after David–Mattila and David's Tb theorems, a sort of the complete theory of Calderón–Zygmund operators in nonhomogeneous setting appeared in Nazarov–Treil–Volberg's [**44**]–[**47**]. These Tb theorems again concerned a general Calderón–Zygmund operator rather than a concrete one (and this turned out to be essential).

Another thing is that Nazarov–Treil–Volberg's Tb theorems were of quantitative nature. Firstly, this gave another (and more streamlined) proof of David's geometric description of removable sets for bounded analytic functions (the problem of Painlevé, or Vitushkin). Secondly, when Tolsa made a decisive move by solving Vitushkin's semi-additivity conjecture in 2001, his proof was based on this quantitative Tb theorem of Nazarov–Treil–Volberg.

The main goal in what follows is to make an exposition of the technique underlying the theory of nonhomogeneous Calderón–Zygmund operators. We choose to illustrate this technique by using it to solve two concrete problems.

The first problem has been already mentioned. It concerns Vitushkin's problem of semiadditivity of analytic capacity. This problem was solved by Tolsa in [**64**]. This article is forcefully not self-contained; it contains many references to the nonhomogeneous Tb theorem of Nazarov–Treil–Volberg. The reader will find below the self-contained proof of this difficult result; especially if the reader will adapt the exposition below to the case $n = 2$. We are saying that because we are dealing, as a rule, with the case $n \geq 3$ in what follows. Thus, we are working with Lipschitz harmonic functions and Lipschitz harmonic capacity (introduced by Mattila and Paramonov) rather than with bounded analytic functions (gradients of Lipschitz harmonic functions) and Vitushkin's analytic capacity. As it is more natural to work with harmonic functions in $\mathbb{R}^n$ than on the plane, the exposition deals with general n. But the reader will be able to extract the "analytic capacity" case, if necessary. In this case our approach differs slightly from Tolsa's approach. This is because only for $n = 2$, the notion of Menger–Melnikov's curvature of measure (see [**37**], [**39**], [**17**], [**32**], [**33**], [**64**]) is related with Lipschitz harmonic capacity, and so we are obliged to work without this fantastically beautiful and powerful tool. One can do this for any dimension n including $n = 2$.

The second problem, which serves as a test of the strength of the theory of nonhomogeneous Calderón–Zygmund operators, is the problem of finding necessary and sufficient conditions for the boundedness of two–weighted Calderón–Zygmund operators. We give such necessary and sufficient conditions in very natural terms, if the operator is the Hilbert transform, and the weights have some mild smoothness. For a certain model Calderón–Zygmund operator, one can give a necessary and sufficient conditions on weights, without assuming anything. We mean here the so-called Martingale transform, which is sometimes considered as a dyadic version of the Hilbert transform.

Notice that the two–weight problem for singular operators seemed to be extremely difficult; adequate tools seemed to be not available. The theory of nonhomogeneous Calderón–Zygmund operators, as we will see below, at least gives considerable hope in understanding such two–weight problems. The second part of our exposition gives several results to this end.

The interest in two–weight problems for singular operators naturally appears from an attempt to understand when the operator in the Hilbert space has an unconditional spectral decomposition. Due to Wermer [71], the following rigidity claim holds: this unconditional spectral decomposition exists for T if and only if $T = S^{-1}NS$, where N is a normal operator, and S is an invertible operator (similarity). This similarity to the normal operator question received much large attention recently for different classes of T. We mention here [4], [40], [31]. If T is a small perturbation of a unitary operator (even a rank one perturbation), then in general the criteria of similarity with a normal operator is more or less totally open. Even if T is a contraction, the relation between the spectral data of U and N is very subtle in general. These kinds of questions immediately become related to two–weight problems for the Cauchy transform, as illustrated by [40]. For example, [40] is based on a remarkable example of Fedor Nazarov, which shows that the Hunt–Muckenhoupt–Wheeden criterion for one–weight boundedness of the Hilbert transform is not applicable in two–weight situations. The reader will find more details in [41], [42] and in Chapter 15, Chapter 16 of this book.

Finally, let us explain the main difficulty of the theory of nonhomogeneous Calderón–Zygmund operators. Roughly speaking the difficulty appears as a result of a certain *degeneracy* in the operator. We can evoke the vague analogy with subellipticity in PDE. In our case, the degeneracy appears not in the kernel of the operator (the kernel is just a classical Calderón–Zygmund kernel) but in underlying measure. To illustrate the kind of difficulty that persistently appears, let us think that we need to estimate the quantity

$$(1.1) \qquad I := |\int_Q \int_R k(x,y)f(x)g(y)\, d\mu(x)\, d\mu(y)| \, .$$

Three possibilities can logically occur: 1) to estimate k in L^∞ (maybe after using some sort of cancellation), and to estimate f in $L^1(\mu)$, g in $L^1(\mu)$; 2) to estimate k in $L^1 L^\infty$ (this is a mixed norm, L^1 in the first variable, L^∞ in the second one), and to estimate f in $L^\infty(\mu)$, g in $L^1(\mu)$; 3) to estimate k in L^1, and to estimate f in $L^\infty(\mu)$, g in L^∞.

In the first case no difficulty appears. We need to bound I in (1.1) by L^2–norms of f, g. And this is not a problem, by the Cauchy inequality

$$\|f\|_{L^1} \le \mu(Q)^{1/2}\|f\|_{L^2}, \quad \|g\|_{L^1} \le \mu(R)^{1/2}\|g\|_{L^2} \, .$$

Suppose we want to repeat something like that in the second case. First of all, L^∞ norm cannot be estimated by the L^2 one. But this is not the difficulty (strangely enough), because in expression I usually f, g are very simple, basically constant functions on Q, R. In this case we have the desired estimates: $\|f\|_{L^\infty} \le \mu(Q)^{-1/2}\|f\|_{L^2}, \|g\|_{L^1} \le \mu(R)^{1/2}\|g\|_{L^2}$. Subsequently, we get the expression $\frac{\mu(R)^{1/2}}{\mu(Q)^{1/2}}$. This is a not so nice expression because measure of a (small) set Q stands in the denominator. For good measures (for example, for Lebesgue measure) we have a control of these "small denominators". But for an arbitrary measure, the denominator can be arbitrarily small, or even vanishing. The only hope is that $R \subset Q$ in all such cases. But this is not so usually. Usually the mutual position of Q, R is quite arbitrary. In the third case there are two small numbers in the denominator. This is even worse. So we are bound for disaster if we reduce the estimate of the operator with kernel k to estimates of sums of expressions of type I. But actually this is exactly the most natural way of estimating Calderón–Zygmund operators. So to avoid this disaster we have to avoid bad mutual positions of Q, R. This goal is attained by considering random decomposition (with respect to random dyadic lattice) of our functions and averaging procedure. This randomness compensates for the degeneracies of the measure because it "smoothens up" the degeneracies, (not in the strictest sense of this word, however). In another context the random dyadic lattice of course already appeared in harmonic analysis, in [22], for example. Decomposition of functions to estimate the Calderón–Zygmund operator is not something new either; see [11]. But the combination of these two ideas is what allows us to win over degeneracies of measures. The machinery of this is represented below; along with two applications (mentioned already) of this technique.

We do not cover here any relations of nonhomogeneous harmonic analysis with the Geometric Measure Theory. The reason is that, of course, in dimensions $n \geq 3$ this relation is not yet established. On the plane the relations are so pervasive and so interesting that discussion of this would require an entire book. Actually, such a book exists. Lecture notes [52] of Pajot cover the geometric part of the topic in great detail. On the other hand, the analytic part of some recent results are only sketched in these lecture notes. In the present exposition we try to fill this gap by presenting the self-contained analytic approach to two chosen topics: semiadditivity of Lipschitz harmonic capacity, and two–weight estimates for certain singular operators. Let us mention recent surveys on this and related topics: [17], [32], [33], [68].

Plan: The first several chapters are devoted to a sort of potential theory with Calderón–Zygmund signed kernels (in this case, Riesz signed kernels). The nonhomogeneous Calderón–Zygmund (CZ) theory is developed starting with Chapter 7. Theorems 7.1 and 8.1 already are the results of the nonhomogeneous Calderón–Zygmund theory. In particular, in Theorem 8.1, the probabilistic language already appears. The proofs of Theorems 7.1 and 8.1 are finished in Chapter 14. The rest of the exposition is devoted to using the same technique in two–weight problems for certain singular operators; this starts in Chapter 15. For the Hilbert transform we give a necessary and sufficient condition of the two–weight boundedness, but only under an extra assumption of "smoothness" of weights. However, in Chapter 22 we give a necessary and sufficient condition of the two–weight boundedness of a three member family of operators, the first member of the family is the Hilbert transform, and the second and the third members are certain simple maximal operators. This necessary and sufficient condition is obtained without any extra assumptions on weights. The same type of criteria can be obtained for singular operators called Martingale transforms. It is known that the Martingale transform is closely related to the Hilbert transform; see, for example, [1], [3]. We do not obtain it here for Martingale Transforms, but the reader can look at [50], [49] for more information.

Acknowledgements. These are the notes of 10 lectures given in May of 2002 at the NSF–supported CBMS conference held at the University of North Carolina at Chapel Hill. I am very grateful to the NSF for the support, and to Professors Joseph Cima and Alex Masterson for the excellent organization. I am very grateful to the stimulating audience. In the first chapters we follow closely the ideas from Tolsa's paper [64] with modifications dictated by a) the lack of complex analysis for higher dimensions, and b) the fact that the magnificent tool of Melnikov–Menger's curvature (see [37], [39], [52]) is "cruelly missing" in higher dimensions.

I am very grateful to Xavier Tolsa for several valuable discussions, which helped to understand better the multidimensional case. I am grateful to Mark Melnikov for his precise and wonderful questions concerning what has become the theory of nonhomogeneous Calderón–Zygmund operators. The mathematics behind the rest of these lectures is due primarily to the group NTV. I was proud to collaborate with Fedor Nazarov (N) and Sergei Treil (T), and I am very grateful to them for this exciting opportunity. The Department of Mathematics of the Michigan State University made this close collaboration possible. I owe my gratitude to the Department. Finally, my deep thanks to the Center of Mathematical Research of the University Autónoma de Barcelona and to Equipe d'Analyse of the Université Paris VI, where these notes were partially prepared.

CHAPTER 2

Preliminaries on Capacities

Let $Lip^1_{\text{loc}}(\mathbb{R}^n)$ be the set of all complex valued locally Lipschitz functions in $\mathbb{R}^n$ (with exponent 1). The fundamental solution Φ for the Laplace equation $\Delta f = 0$ in $\mathbb{R}^n$ is defined by $\Phi(x) = -\frac{a_n}{|x|^{n-2}}$, $n \geq 3$, where $a_n > 0$ is a constant. $\Phi(x) = -\frac{1}{2\pi} \log \frac{1}{|x|}$ for $n = 2$.

We now introduce the class of admissible functions for the definition of Lipschitz harmonic capacity. For a compact set E in $\mathbb{R}^n$, set

$$L(E, 1) := \{f \in Lip^1_{\text{loc}}(\mathbb{R}^n) : \text{supp}(\Delta f) \subset E, \|\nabla f\|_\infty \leq 1, \nabla f(\infty) = 0\},$$

where $\text{supp}(\Delta f)$ is the support of the distribution Δf.

We shall consider functions modulo constants in $L(E, 1)$, meaning that we shall write $f = g$ for $f, g \in L(E, 1)$ if $f - g$ is constant. Note that each function $f \in L(E, 1)$ is harmonic in $\mathbb{R}^n \setminus E$ and $f = \Phi * \Delta f + \text{constant}$.

The Lipschitz harmonic capacity of the set E is defined by

$$\gamma(E) := \sup\{|\langle \Delta f, 1 \rangle| : f \in L(E, 1)\},$$

where (as usual) $\langle S, \varphi \rangle$ means the action of the distribution with compact support on a smooth test function.

Letting $\alpha(n-1)$ be the volume of the unit ball in $\mathbb{R}^{n-1}$, we define the $(n-1)$-dimensional Hausdorff measure for a subset E in $\mathbb{R}^n$ by

$$\mathcal{H}^{n-1}(E) := \lim_{\delta \to 0} \inf\{\sum_{i=1}^\infty \alpha(n-1)r_i^{n-1} : E \subset \cup_{i=1}^\infty B(x_i, r_i), \, r_i \leq \delta\}.$$

Then the restriction of $\mathcal{H}^{n-1}$ to sufficiently regular hypersurfaces gives the surface measure.

A set E in $\mathbb{R}^n$ is called removable for Lipschitz harmonic functions if for each domain D in $\mathbb{R}^n$ every locally Lipschitz function $f : D \to \mathbb{C}$ which is harmonic in $D \setminus E$ is harmonic in D. It is proved in [**35**] that

A set E is removable for Lipschitz harmonic functions if and only if $\gamma(E) = 0$.

A set E in $\mathbb{R}^n$ is called removable for subharmonic Lipschitz harmonic functions if for each domain D in $\mathbb{R}^n$ every locally Lipschitz function $f : D \to \mathbb{C}$ which is harmonic in $D \setminus E$ and subharmonic in D is harmonic in D.

Such sets also can be characterized using a natural capacity. Let us consider

$$\gamma_+(E) := \sup\{|\langle \Delta f, 1 \rangle| : f \in L(E, 1), \Delta f = \mu \in M_+(E)\},$$

where $M_+(E)$ is the family of positive Borel measures on E. Similarly to [**35**], one can see that

A set E is removable for subharmonic Lipschitz harmonic functions

if and only if $\gamma_+(E) = 0$.

By definition,

$$(2.1) \qquad\qquad\qquad \gamma_+(E) \leq \gamma(E) \,.$$

Or, obviously, sets removable for Lipschitz harmonic functions are removable for subharmonic Lipschitz harmonic functions. We will prove here (in the next 125 pages) that the converse is true. Moreover, we prove

THEOREM 2.1. *There exists a constant A depending only on the dimension n such that*

$$(2.2) \qquad\qquad\qquad \gamma(E) \leq A\gamma_+(E) \,.$$

In the case $n = 2$, the attention of many mathematicians was attracted to the comparison of analytic capacity and "positive analytic capacity". The capacities γ, γ_+ are very close relatives of these classical objects, which, abusing the conventional notation, we will call Γ, Γ_+ (usually analytic capacity is called γ, here we use the symbol Γ).

Definition. Let E be a compact set in $\mathbb{C}$.

$$\Gamma(E) := \sup\{\lim_{z \to \infty} |z\, f(z)| : f \in \text{Hol}(\mathbb{C} \setminus E), |f(z)| \leq 1 \ \forall z \in \mathbb{C} \setminus E, \ f(\infty) = 0\} \,,$$

$$\Gamma_+(E) := \sup\{\lim_{z \to \infty} |z\, f(z)| : f(z) = \int \frac{d\mu(\zeta)}{z - \zeta}, \mu \in M_+(E), |f(z)| \leq 1 \ \forall z \in \mathbb{C} \setminus E\} \,.$$

By definition,

$$(2.3) \qquad\qquad\qquad \Gamma_+(E) \leq \Gamma(E) \,.$$

Parallel to Theorem 2.1, we will prove here the following great result of X. Tolsa [64] that the converse to (2.3) is true. This result brought the solution of a semiadditivity of analytic capacity conjecture of Vitushkin and the conjecture of Melnikov that every set of positive Γ supports positive measure of order 1 (meaning that $\mu(B(x,r)) \leq r, \ \forall x \in \mathbb{C}, r > 0$) and finite Menger's curvature; see [37].

The proof in [64] was obliged to refer many times to the theory of nonhomogeneous Calderón–Zygmund operators, especially to results from Nazarov–Treil–Volberg [46], [47]. Our exposition, in this respect, can be considered not only as proving Theorem 2.1, but also it might turn out to be the first self-contained exposition of the proof of Tolsa's Theorem 2.2.

THEOREM 2.2. *There exists a constant A such that*

$$(2.4) \qquad\qquad\qquad \Gamma(E) \leq A\,\Gamma_+(E) \,.$$

The reader will notice soon that there is a great similarity between γ and Γ and between γ_+ and Γ_+. One can consider Theorem 2.1 to be a correct multidimensional analog of Theorem 2.2.

Exploring similarity of definitions, one can see that γ_+ is defined by imposing conditions on potential $k \star \mu$, where k is a vector valued Riesz kernel, and Γ_+ is defined by imposing conditions on potential $K \star \mu$, where $K(z, \zeta) = \frac{1}{z-\zeta}$ is the (complex-valued) Cauchy kernel. In both cases, the kernels are Calderón–Zygmund (CZ) kernels of order $n - 1$, when we are in $\mathbb{R}^n$, $n \geq 2$.

There are also some differences. In one respect, Γ theory is more difficult than γ theory (we compare them, of course, only for $n = 2$), because the restriction on potential $k \star \mu$ is imposed a.e. in $\mathbb{R}^2$ for γ theory (it follows from the equivalent definitions of Lipschitz harmonic capacities) and only in $\mathbb{R}^2 \setminus E$ for Γ theory. This difference is not as trivial as one could think. For example, if we introduce

$$C_+(E) := \sup\{|\langle \Delta f, 1\rangle| : |\nabla f(x)| \leq 1, \forall x \in \mathbb{R}^n \setminus E, \Delta f = \mu \in M_+(E)\},$$

we hit the following unsolved

Problem. Is $C_+(E) \asymp \gamma_+(E)$ with constants independent of E, $n \geq 3$?

In principle, this could have been a problem for $n = 2$ as well. The reason why the difference between a.e. restriction and "outside of E restriction" turns out to be inessential for $n = 2$ is the following. Fortunately, to prove Theorem 2.2 it is sufficient to prove it (with constant independent of E) only for E from a very special class, namely, only for E that are a finite union of smooth curves on the plane (and these are undoubtedly the sets of zero planar Lebesgue measure). For $n \geq 3$ this remark may not work, and the problem remains. Of course, all of this is a problem only if Lebesgue measure of E is positive.

We just explained why Γ theory is worse than γ theory. But in another respect, γ theory is worse than Γ theory, because for the Cauchy potential (dealing with Γ) we can use Complex Analysis, and for Riesz potentials (dealing with γ) it is unavailable (at least for $n \geq 3$). This creates two serious complications for adapting Tolsa's theorem for the proof of Theorem 2.1:

1) Simple localization estimates for the potential should be obtained without Complex Analysis. This is done in Chapter 3.

2) Suitable analogs of Menger's curvature and all magic formulae are "cruelly missing" by the expression of Guy David in higher dimensions (see [**20**]). And these tools were essentially used by Tolsa in proving Theorem 2.2. We have to circumvent this difficulty too. This is done in Chapter 5.

We want to discuss a bit the genesis of the problems. The problems have been in the focus of attention of many analysts recently. There is a fantastically beautiful geometric side of the story involving P. Jones' traveling salesman problem and its analogs, but we do not touch this here at all! We refer the reader to the concise introduction to the geometric side of the story in the book of H. Pajot [**52**].

The analytical part of the story was started by the article of M. Christ [**5**]. He proved Theorem 2.2 for so-called Ahlfors regular sets. The struggle with extra regularity brought the special Tb theorem of G. David and P. Mattila [**18**] and David [**19**] and the series of works by Tolsa. At the same time, general approach to nonhomogeneous Calderón–Zygmund theory in Nazarov–Treil–Volberg's [**43**]-[**47**] turned out to be possible, and, as we will see, useful. This approach does not use extra regularity of measures. The reader can have a brief but very good account of this in reviews of David [**17**] and Mattila [**32**], [**33**]. Briefly speaking, the outcome of the approach developed in [**43**]-[**47**] is that estimates are simpler and the statements are more flexible and give rise to quite general nonhomogeneous Tb theorems with *bounds*. These bounds were used by Tolsa in [**64**] to prove Theorem 2.2, which crowned all of the previous efforts.

As we have already said, our goal here will be to give simultaneously the self-contained proof of Theorem 2.2 and of its Lipschitz harmonic analog in Theorem 2.1. One of the byproducts of all of this is that on the plane all of our capacities are equivalent:

$$\gamma \asymp \Gamma \asymp \gamma_+ \asymp \Gamma_+.$$

We concentrate our efforts on the proof of Theorem 2.1. Why Theorem 2.2 is simultaneously proved, and what other capacities can be included in the scheme, is discussed in Section 5.5.

Notice that some particular examples of Theorem 2.1 were considered in [18], [36].

CHAPTER 3

Localization of Newton and Riesz Potentials

3.1. Localization lemmas

Recall that Newton's kernel is defined by $\Phi(x) = -\frac{a_n}{|x|^{n-2}}$, $n \geq 3$, where $a_n > 0$ is a constant, $\Phi(x) = -\frac{1}{2\pi} \log \frac{1}{|x|}$ for $n = 2$. Let $\Psi = \nabla \Phi$; this is a Riesz kernel (actually it is a vector Riesz kernel consisting of n Riesz kernels Ψ^i which are $\partial_i \Phi := \frac{\partial}{\partial x_i} \Phi$). These are Calderón–Zygmund (CZ) kernels having singularity of order $n - 1$. Sometimes we use the notation

$$W^S(x) = (\Phi * S)(x), \; V^S(x) = (\Psi * S)(x),$$

for the Newton and Riesz potentials of distribution S, respectively.

We always consider E in $B_0 := B(0, 1)$. The letter S is reserved for distributions with compact support, dx denotes Lebesgue measure in $\mathbb{R}^n$. The letter A denotes any absolute constant or a constant that depends only on n; $|\cdot|$ stands for the norm in $\mathbb{C}^n$.

LEMMA 3.1. *Let S be supported on E and $\Phi * S \in L(E, 1)$. Let $B = B(x, r)$ be a ball, and let $\varphi \in C_0^\infty(2B)$ be such that $\nabla^2 \varphi \leq \frac{A}{r^2}$. Then*

$$(3.1) \qquad |\langle S, \varphi \rangle| \leq A \, r^{n-1} \, .$$

Proof. Let $f = \Phi * S \in L(E, 1)$. Then $|f(y) - f(x)| \leq 2r \, \forall y \in B(x, 2r)$. Now $|\langle S, \varphi \rangle| = |\langle \Delta f, \varphi \rangle| = |\langle \Delta(f - f(x)), \varphi \rangle| = |\langle (f - f(x)), \Delta \varphi \rangle| \leq \frac{A r}{r^2} \text{vol}(B(x, r)) \leq A \, r^{n-1}$. $\qquad \square$

Consider the following situation. Distribution S is supported on $E \subset B := B(x, r)$, $\Phi * S \in L(E, 1)$.

LEMMA 3.2. *Let φ be from $C^\infty(2B)$, and*

$$\|\varphi\|_{L^\infty(2B)} \leq Dr, \; \|\nabla \varphi\|_{L^\infty(2B)} \leq D, \; \|\nabla^2 \varphi\|_{L^\infty(2B)} \leq \frac{D}{r} \, .$$

Then

$$(3.2) \qquad |\langle S, \varphi \rangle| \leq ADr^n \, ,$$

and moreover

$$(3.3) \qquad |\langle S, \varphi \rangle| \leq ADr \, \gamma(E) \, .$$

Proof. Consider any $g \in C_0^\infty(2B)$ such that $g = 1$ on B, $0 \leq g \leq 1$, $\nabla^2 g \leq \frac{A}{r^2}$. Let $f = \Phi * S \in L(E, 1)$. Then $|f(y) - f(x)| \leq 2r \, \forall y \in B(x, 2r)$.

Put $F := \varphi g(f - f(x))$. Then

$$\Delta F = \varphi g S + \Delta(\varphi g)(f - f(x)) + \langle \nabla(\varphi g), V * S \rangle =: I + II + III \, .$$

Notice that II and III are bounded functions supported on $2B$. Hence $F = \Phi * I + \Phi * II dx + \Phi * III dx + H$, where H is harmonic in $\mathbb{R}^n$ and vanishes at infinity. So $H = 0$. Therefore,

$$(3.4) \qquad \Psi * I = \nabla F - \Psi * II dx - \Psi * III dx.$$

Let us estimate functions II and III. $\|II\|_{L^\infty(2B)} \leq \|\nabla^2 \varphi\| \|g\| \|f - f(x)\| + \|\nabla \varphi\| \|\nabla g\| \|f - f(x)\| + \|\varphi\| \|\nabla^2 g\| \|f - f(x)\| \leq A\frac{D}{r}r + AD\frac{A}{r}r + ADr\frac{A}{r^2}r \leq AD$. Similarly, $\|III\|_{L^\infty(2B)} \leq AD$. As II and III are supported on $2B$, we now get

$$\|\Psi * II dx\|_{L^\infty(2B)} \leq ADr, \quad \|\Psi * III dx\|_{L^\infty(2B)} \leq ADr.$$

So by the maximum value theorem for harmonic functions,

$$(3.5) \qquad |\Psi * II dx|(y) \leq ADr, \quad |\Psi * III dx|(y) \leq ADr \; \forall y \in \mathbb{R}^n.$$

The first term in (3.4) is zero outside of $2B$. Inside

$$(3.6) \qquad |\nabla F(y)| \leq |g \nabla \varphi (f - f(x)) + |\varphi \nabla g(f - f(x)) + |\varphi g \, \Psi * S| \leq ADr.$$

Gathering (3.4), (3.5), (3.6), we get

$$(3.7) \qquad |(\Psi * \varphi S)(y)| = |(\Psi * \varphi g S)(y)| \leq ADr \; \forall y \in \mathbb{R}^n.$$

Consider now a new distribution $S_\varphi := \varphi S / ADr$. Then (3.7) shows that

$$S_\varphi \in L(E, 1).$$

Apply Lemma 3.1 to S_φ. Then

$$|\langle S_\varphi, g \rangle| \leq A r^{n-1}.$$

This implies

$$|\langle S, \varphi \rangle| = |\langle S, \varphi g \rangle| = ADr |\langle S_\varphi, g \rangle| \leq ADr^n.$$

This is (3.2). But actually, the definition of $\gamma(E)$ and (3.7) shows that

$$|\langle S_\varphi, 1 \rangle| \leq \gamma(E).$$

But this is the same as

$$|\langle S, \varphi \rangle| = ADr |\langle S_\varphi, 1 \rangle| \leq ADr\,\gamma(E),$$

which is (3.3).

$\square$

Recall that we consider the following situation. Distribution S is supported on $E \subset B := B(x, r)$, $\Phi * S \in L(E, 1)$. Consider Lemma 3.2 with a special φ. Fix a $z \in \mathbb{R}^n \setminus 3B$ and put

$$\varphi_z(y) := \Phi(z - y) - \Phi(z - x).$$

Notice that it satisfies the assumptions of Lemma 3.2 with $D = \frac{1}{\text{dist}(z,E)^{n-1}}$. Thus, we get the following corollary.

COROLLARY 3.3. *If distribution S is supported on $E \subset B := B(x, r)$ and $\Phi * S \in L(E, 1)$, then $|(\Phi * S)(z) - \langle S, 1 \rangle \Phi(z - x)| = |\langle S, \varphi_z(y) \rangle|$ and therefore*

$$(3.8) \qquad |(\Phi * S)(z) - \langle S, 1 \rangle \Phi(z - x)| \leq \frac{A r \gamma(E)}{\text{dist}(z, E)^{n-1}}.$$

In particular, if $\langle S, 1 \rangle = 0$, we get

$$(3.9) \qquad |(\Phi * S)(z)| \leq \frac{A\, r\, \gamma(E)}{\text{dist}(z, E)^{n-1}}\,,$$

and

$$(3.10) \qquad |(\Psi * S)(z)| \leq \frac{A\, r\, \gamma(E)}{\text{dist}(z, E)^n}\,.$$

3.2. A building block for the construction of special measures

Consider again the following situation. Distribution S is supported on $E \subset B := B(x, r)$, $\Phi * S \in L(E, 1)$. Let us modify distribution S. First of all, consider measure μ, which is a surface measure on the sphere of radius $R = c_n \gamma(E)^{\frac{1}{n-1}}$ centered at x. The constant c_n is chosen in such a way that $R \leq r/n$ and $\|\mu\| \asymp \gamma(E)$, where the constants of comparison depend only on dimension n. Let

$$\hat{S} = \langle S, 1 \rangle \frac{\mu}{\|\mu\|}\,.$$

Of course, complex measure $\hat{S}$ satisfies

$$(3.11) \qquad \langle \hat{S}, 1 \rangle = \langle S, 1 \rangle\,.$$

Also, (3.8) can be rewritten as

$$(3.12) \qquad |(\Phi * S)(z) - (\Phi * \hat{S})(z)| \leq \frac{A\, r\, \gamma(E)}{\text{dist}(z, E)^{n-1}} \quad \forall z \in \mathbb{R}^n \setminus 3B\,.$$

3.3. Localization on special cubes

Let $E \subset B_0 := B(0, 1)$. We consider the family of cubes with the following properties:

$$(3.13) \qquad \gamma_+(\cup_{i=1}^N Q_i) \leq C_0 \gamma_+(E)\,,$$

$$(3.14) \qquad \sum_{i=1}^N \gamma_+(2Q_i \cap E) \leq C_1 \gamma_+(E)\,,$$

$$(3.15) \qquad \text{diam}\, Q_i \leq \frac{1}{10}\, \text{diam}\, E\,.$$

Cubes $\{Q_i\}$ will have bounded multiplicity of overlapping. Constants C_0, C_1 depend only on the dimension n. Such a family of cubes will be constructed below. Unlike other constants depending only on the dimension, we keep special names for these constants because of their special part in the proof.

Consider $\{g_i\}_{i=1}^N$, $g_i \in C_0^\infty(2Q_i)$, $0 \leq g_i \leq 1$ on $\mathbb{R}^n$, $|\nabla^2 g_i| \leq \frac{A}{\ell(Q_i)^2}$. In addition,

$$(3.16) \qquad \sum_{i=1}^N g_i = 1 \quad \text{on} \quad \cup_{i=1}^N Q_i\,.$$

Let $\Phi * S \in L(E, 1)$, $E \subset B_0 := B(0, 1)$.

LEMMA 3.4. *There exists $A < \infty$ (depending only on the dimension) such that $\frac{1}{A} \Phi * g_i S \in L(E, 1)$. Henceforth, $\frac{1}{A} \Phi * g_i S \in L((2Q_i \cap E), 1)$.*

Proof. Fix i. Let x be the center of Q_i. Put $f := \Phi * S$ and $F := g_i(f - f(x))$. Then $\Delta F = g_i S + \Delta g_i(f - f(x)) + \langle \nabla g_i, \Psi * S \rangle =: I + II + III$. Hence

$$F = \Phi * g_i S + \Phi * II dx + \Phi * III dx + H \,,$$

where H is harmonic in $\mathbb{R}^n$. Notice that II, III are supported on $2Q_i$, bounded and $\|II\|_{L^\infty} \leq \frac{A}{\ell(Q_i)}, \|III\|_{L^\infty} \leq \frac{A}{\ell(Q_i)}$. Therefore, the first three terms on the right hand side of the previous equality vanish at infinity. Its left hand side F vanishes outside $2Q_i$, so H vanishes at infinity, and this means that $H = 0$. Now we can write

$$(3.17) \qquad \Psi * g_i S = \nabla F - \Psi * II dx - \Psi * III dx \,.$$

We already saw that II, III are such that $\|II\|_{L^\infty} \leq \frac{A}{\ell(Q_i)}, \|III\|_{L^\infty} \leq \frac{A}{\ell(Q_i)}$, and they are supported by $2Q_i$. In particular, $|(\Psi * II dx)(y)| \leq A$, $|(\Psi * III dx)(y)| \leq A$, $\forall y \in 2Q_i$, and then, by the maximal principle,

$$(3.18) \qquad |(\Psi * II dx)(y)| \leq A, \ |(\Psi * III dx)(y)| \leq A, \ \forall y \in \mathbb{R}^n \,.$$

The first term ∇F in (5.62) is also uniformly bounded. In fact, $\nabla F = \nabla g_i(f - f(x)) + g_i \Psi * S$. The second term here is bounded by the assumption: $\Phi * S \in L(E, 1)$. The first term is bounded because we have to consider it only on $2Q_i$ (outside it is zero) and on $2Q_i$ $|f - f(x)| \leq A\ell(Q_i)$ (again by the fact that $\Phi * S \in L(E, 1)$).

Finally, (3.17), (3.18) and the uniform boundedness of ∇F imply that $\Psi * g_i S$ is bounded in $\mathbb{R}^n$. The lemma is proved. $\qquad \square$

Now we obviously have

COROLLARY 3.5. *Let $E \subset B(0, 1)$, $\Phi * S \in L(E, 1)$, $\{Q_i\}$ satisfy (3.13)-(3.15). Then*

$$(3.19) \qquad |\langle S, g_i \rangle| \leq A \gamma(2Q_i \cap E) \,.$$

3.4. Modification of distribution S. Construction of auxiliary measures

Let us recall that we have potentials $W = \Phi*, V = \Psi*$. Let us combine Lemma 3.2 and Lemma 3.4 to build special measures μ, ν starting with distribution S such that

$$\Phi * S \in L(E, 1) \,.$$

We fix $i = 1, \ldots, N$ and consider the distribution $g_i S$ which is supported on $2Q_i$. This set plays the role of $2B(x, r)$ from Lemma 3.2, and $g_i S$ plays the role of S from this lemma. We use the construction from Section 3.2 with $g_i S$ playing the role of S. We already proved (see (3.12)) that

$$(3.20) \qquad |W^{g_i S}(z) - W^{\widehat{g_i S}}(z)| \leq \frac{A \ell(Q_i) \gamma(2Q_i \cap E)}{\text{dist}(z, Q_i)^{n-1}} \ \forall z \in \mathbb{R}^n \setminus 3Q_i \,,$$

and consequently

$$(3.21) \qquad |V^{g_i S}(z) - V^{\widehat{g_i S}}(z)| \leq \frac{A \ell(Q_i) \gamma(2Q_i \cap E)}{\text{dist}(z, Q_i)^n} \ \forall z \in \mathbb{R}^n \setminus 3Q_i \,.$$

Also by Lemma 3.4,

$$(3.22) \qquad |V^{g_i S}(z)| \leq A \ \forall z \in \mathbb{R}^n \,.$$

Recall that we have a sphere $\Sigma_i(x_i, R_i)$ lying inside Q_i centered at the center x_i of Q_i, such that its surface measure $\mathcal{H}^{n-1}(\Sigma_i(x_i, R_i))$ is comparable (with constants depending only on the dimension) to $\gamma(2Q_i \cap E)$, and we introduced measures

$$\mu_i := \mathcal{H}^{n-1}|\Sigma_i(x_i, R_i) \quad \mu := \sum \mu_i .$$

Now let us replace the distribution $S = \sum g_i S$ by complex measure

(3.23) $$\nu := \sum \widehat{g_i S} = \sum \langle S, g_i \rangle \frac{\mu_i}{\|\mu_i\|} .$$

Now (3.19) gives that

(3.24) $$\|\frac{d\nu}{d\mu}\|_\infty \leq \hat{A} .$$

We know that $|V^S| \leq 1$ in $\mathbb{R}^n$. Our next goal is to see what kind of boundedness we have for the potential of the modified measure ν. We will "almost prove" that

$$\int |V^\nu| d\mu \leq A\|\mu\| .$$

3.5. Ahlfors balls

Lemma 3.1 proves (we should choose $r = \operatorname{diam}(E)$ and $\varphi = 1$ on E in Lemma 3.1) the existence of the constant $A(n) < \infty$ such that

(3.25) $$\gamma(E) \leq A_1(n)(\operatorname{diam}(E))^{n-1} .$$

Recall that there exists a positive finite constant $A_2(n)$ such that

(3.26) $$A_2^{-1}\|\mu_i\| \leq \gamma(2Q_i \cap E) \leq A_2\|\mu_i\| .$$

Let us use these constants $A_1(n), A_2(n)$ along with C_1 from (3.14) to introduce

(3.27) $$C_2 = 100^{n-1} A_1(n)A_2(n) C_1 .$$

Let us also use from now on the notation

$$F := \cup_{i=1}^N Q_i .$$

We have $\mu(F) = \sum \|\mu_i\| \leq A_2 \sum \gamma(2Q_i \cap E)$, and so

$$\mu(F) \leq C_1 A_2 \gamma(E) \leq C_1 A_2 \gamma(F) \leq C_1 A_1 A_2 \operatorname{diam}(F)^{n-1} .$$

This immediately implies

(3.28) $$\forall x \in F \,\forall R > \frac{\operatorname{diam} F}{100}, \quad \mu(B(x, R)) \leq C_2 R^{n-1} .$$

Now we are ready to introduce the notion of a non-Ahlfors ball. The ball $B(x, R), x \in F$ is called non-Ahlfors if $\mu(B(x, r)) > C_2 R^{n-1}$. Otherwise it is called an Ahlfors ball. We see that every ball centered at $x \in F$ is an Ahlfors ball if its radius is larger than $\frac{\operatorname{diam} F}{100}$. Denote by $R(x)$ the supremum of all non-Ahlfors balls centered at x. Ahlfors points are those for which $R(x) = 0$. We know that $\forall x \, R(x) \leq \frac{\operatorname{diam} F}{100}$.

Denote

$$H_0 := \cup_{x \in F} B(x, R(x)) .$$

By Vitali's covering lemma choose disjoint $B(x_k, R(x_k))$ and put

$$H = \cup_k B(x_k, 3R(x_k)) .$$

Then all non-Ahlfors balls are contained in H and

$$(3.29) \qquad \text{dist}(x, F \setminus H) \geq R(x), \ \forall x \in H \cap F,$$

$$(3.30) \qquad \sum_k R(x_k)^{n-1} \leq \frac{1}{C_2} \mu(F).$$

3.6. The principal estimate for auxiliary measures

Let us set

$$V_*^\nu(y) := \sup_{\varepsilon > 0} |(\Psi * \chi_{\mathbb{R}^n \setminus B(y,\varepsilon)} d\nu)(y)|.$$

Following [64], we prove

THEOREM 3.6.

$$(3.31) \qquad \int_{F \setminus H} V_*^\nu(y) d\mu(y) \leq C(C_2)\mu(F).$$

Proof. Let ψ be a bell-like function with compact support, $\psi_\varepsilon(x) = \frac{1}{\varepsilon^n} \psi(\frac{x}{\varepsilon})$, and

$$U_\varepsilon^\sigma := \psi_\varepsilon * V^\sigma, \ U_*^\sigma = \sup_\varepsilon |U_\varepsilon^\sigma|.$$

The kernel of U_ε is $k_\varepsilon := \psi_\varepsilon * \frac{x}{|x|^n}$ (here x is a vector $(x_1, \ldots, x_n)$), and so

$$(3.32) \qquad k_\varepsilon(x) = \frac{x}{|x|^n}, \text{ if } |x| > \varepsilon \text{ and } |k_\varepsilon(x)| \leq \frac{A}{\varepsilon^{n-1}} \text{ otherwise}.$$

In particular, given a complex measure ν, we introduce the maximal function

$$M\sigma(x) := \sup_r \frac{|\sigma|(B(x,r))}{r^{n-1}}$$

and write

$$(3.33) \qquad |U_\varepsilon^\sigma(x) - V_\varepsilon^\sigma(x)| \leq AM\sigma(x).$$

Remark. Maximal function M will be used only for measures (and, of course, never for distributions).

LEMMA 3.7.

$$(3.34) \qquad U_*^{g_i S - \widehat{g_i S}}(z) \leq \frac{A\ell(Q_i)\gamma(2Q_i \cap E)}{\text{dist}(z, 2Q_i)^n} \ \forall z \in \mathbb{R}^n \setminus 4Q_i.$$

Let us postpone the proof of Lemma 3.7 and finish now the proof of Theorem 3.6. We start with

$$\int_{F \setminus H} V_*^\nu d\mu = \int_{F \setminus H} (U_*^\nu - V_*^\nu) d\mu + \int_{F \setminus H} U_*^\nu d\mu =: I + II.$$

Let us use the definition of H, (3.27) together with (3.33) and (3.24) to write

$$(3.35) \qquad |I| \leq A \int_{F \setminus H} M\nu d\mu \leq A\hat{A} \int_{F \setminus H} M\mu d\mu \leq A\hat{A}C_2\mu(F).$$

To estimate II we write

$$(3.36) \qquad \int_{F \setminus H} U_*^\nu d\mu \leq \int_{F \setminus H} U_*^S d\mu + \int_{F \setminus H} U_*^{S-\nu} d\mu.$$

Recall that $|V^S(y)| \le A$ everywhere, hence its convolution with ψ_ε is bounded, and so $U_*^S = \sup_\varepsilon |\psi_\varepsilon * V^S|$ is bounded everywhere by a constant A. So the first integral in (3.36) is bounded by $A\mu(F)$. To estimate,

$$\int_{F \setminus H} U_*^{S-\nu} d\mu \le \sum_{i=1}^N \int_{F \setminus H} U_*^{g_i S - \widehat{g_i S}} d\mu$$

$$\sum_{i=1}^N \left(\int_{4Q_i} \cdots + \int_{F \setminus (4Q_i \cup H)} \cdots \right) := \sum_{i=1}^N (T_{i1} + T_{i2}).$$

Let $\alpha_i := g_i S - \widehat{g_i S}$. We want to estimate $|(\psi_\varepsilon * V^{\alpha_i})(z)|, z \in 4Q_i$. We know that functions $V^{g_i S}$ are uniformly bounded. Just use (3.22). Let us prove now that

(3.37) $$\|V^{\widehat{g_i S}}\|_{L^\infty(\mathbb{R}^n)} \le A.$$

Let Σ be a sphere of radius R centered at x on which measure $\widehat{g_i S}$ is concentrated. Let σ denote the normalized surface measure on Σ. It is easy to see that

$$V^\sigma(z) = 0 \ |z - x| < R, \ V^\sigma(z) = \frac{z - x}{|z - x|^n}, \ |z - x| > R.$$

If we use (3.19),(3.26), we get $|\langle S, g_i \rangle| \le A(\gamma(2Q_i \cap E))^{n-1} \le A R^{n-1}$. But the calculation of V^σ above gives $|V^\sigma| \le \frac{A}{R^{n-1}}$ almost everywhere. Then using $|V^{\widehat{g_i S}}(z)| = |\langle S, g_i \rangle||V^\sigma(z)|$, we get (3.37).

In particular,

(3.38) $$T_{i1} \le A\mu(4Q_i).$$

To estimate T_{i2} we use Lemma 3.7. Let N be the least integer such that $B_N := (4^{N+1}Q_i \setminus 4^N Q_i) \setminus H \ne \emptyset$. Then

$$T_{i2} \le A\ell(Q_i)\mu(Q_i) \sum_{k=N}^\infty \int_{B_N} \frac{d\mu(z)}{\text{dist}(z, 2Q_i)^n}.$$

We continue by

$$T_{i2} \le A\ell(Q_i)\mu(Q_i) \sum_{k=N}^\infty \frac{\mu(4^{k+1}Q_i)}{4^{kn}\ell(Q_i)^n}.$$

Now let $z_0 \in B_N$. Then for $k \ge N$ we have

$$\mu(4^{k+1}Q_i) \le \mu(B(z_0, A\ell(4^{k+1}Q_i))) \le A 4^{k(n-1)}\ell(Q_i)^{n-1}.$$

We used here the fact that $z_0 \notin H$. We can continue the estimate:

(3.39) $$T_{i2} \le A\ell(Q_i)\mu(Q_i)\frac{1}{\ell(4^N Q_i)} \le A\mu(Q_i).$$

Combining (3.38), (3.39) one obtains

(3.40) $$\int_{F \setminus H} U_*^{S-\nu} d\mu \le A \sum_{i=1}^N \mu(4Q_i).$$

Combining with (3.35), (3.36), (3.38) and using the fact of finite overlapping of $4Q_i$, one gets

$$\int_{F \setminus H} V_*^\nu d\mu \le A\mu(F).$$

Theorem 3.6 is completely proved. $\qquad\square$

Let us prove now Lemma 3.7.

Proof. Fix $z \in \mathbb{R}^n \setminus 4Q_i$ and let $\varepsilon \leq \frac{\mathrm{dist}(z,2Q_i)}{2}$. Then for any $y \in B(z,\varepsilon)$, we have $\mathrm{dist}(y, 2Q_i) \asymp \mathrm{dist}(z, 2Q_i)$, and every such y is outside of $3Q_i$. Therefore, for all $y \in B(z,\varepsilon)$ we have (3.21):

$$(3.41) \qquad |V^{g_i S}(y) - V^{\widehat{g_i S}}(y)| \leq \frac{A\,\ell(Q_i)\gamma(2Q_i \cap E)}{\mathrm{dist}(z, Q_i)^n}.$$

Making the convolution with ψ_ε, we get

$$(3.42) \qquad |U_\varepsilon^{g_i S - \widehat{g_i S}}(z)| \leq \frac{A\,\ell(Q_i)\gamma(2Q_i \cap E)}{\mathrm{dist}(z, Q_i)^n}.$$

Now fix $z \in \mathbb{R}^n \setminus 4Q_i$ and let $\varepsilon > \frac{\mathrm{dist}(z,2Q_i)}{2} > \ell(Q_i)$. Put $\alpha_i := g_i S - \widehat{g_i S}$. Then

$$U_\varepsilon^{g_i S - \widehat{g_i S}} = V^{\psi_\varepsilon * \alpha_i \, dx}.$$

Then it is easy to see (using that $V = \Psi *$ and $|\Psi(x-y)| \leq \frac{A}{|x-y|^{n-1}}$) that

$$(3.43) \qquad |U_\varepsilon^{\alpha_i}(z)| \leq A\,\|\psi_\varepsilon * \alpha_i\|_{L^\infty}\, \mathrm{diam}(\mathrm{supp}(\psi_\varepsilon * \alpha_i)).$$

But $\mathrm{diam}(\mathrm{supp}(\psi_\varepsilon * \alpha_i)) \asymp \varepsilon$ as $\varepsilon > \ell(Q_i)$.

On the other hand (we denote by x_i the center of Q_i),

$$(\psi_\varepsilon * \alpha_i)(w) = \langle \psi_\varepsilon(w - \xi), \alpha_i(\xi) \rangle = \langle \psi_\varepsilon(w - \xi) - \psi_\varepsilon(w - x_i), \alpha_i(\xi) \rangle$$
$$= \langle (\psi_\varepsilon(w - \xi) - \psi_\varepsilon(w - x_i))\eta_i(\xi), \alpha_i(\xi) \rangle,$$

where $\eta_i \in C_0^\infty(3Q_i), \eta_i = 1$ on $2Q_i$ (recall that $\mathrm{supp}(\alpha_i) \subset 2Q_i$, $|\nabla \eta_i| \leq \frac{A}{\ell(Q_i)}$). We are almost in a position to use Lemma 3.2, namely (3.3) with $D = \frac{1}{\varepsilon^{n+1}}$. In fact, actually, Lemma 3.4 and relation (3.37) show that for a certain constant A (which depends only on the dimension), $\frac{1}{A}\Phi * \alpha_i \in L(E, 1)$. So α_i plays the part of S in Lemma 3.2. To employ Lemma 3.2 we need to verify that $\varphi(\xi) := (\psi_\varepsilon(w - \xi) - \psi_\varepsilon(x_i - \xi))\eta_i(\xi)$ satisfies the assumptions of this lemma with $r \asymp \ell(Q_i), D = \frac{1}{\varepsilon^{n+1}}$. So let us verify that

$$(3.44) \qquad \|\varphi\|_{L^\infty} \leq \frac{A\ell(Q_i)}{\varepsilon^{n+1}}, \ \|\nabla\varphi\|_{L^\infty} \leq \frac{A}{\varepsilon^{n+1}}, \ \|\nabla^2\varphi\|_{L^\infty} \leq \frac{A}{\ell(Q_i)\varepsilon^{n+1}}.$$

But it is easy to see that

$$|\varphi(\xi)| \leq \frac{A}{\varepsilon^{n+1}}\|\psi\|_{L^\infty}\|\eta_i\|_{L^\infty} \leq \frac{A\ell(Q_i)}{\varepsilon^{n+1}},$$
$$|\nabla\varphi(\xi)| \leq |\psi_\varepsilon(w - \xi) - \psi_\varepsilon(w - x_i)||\nabla\eta_i(\xi)| + |\nabla\psi_\varepsilon(w - \xi)||\eta_i(\xi)|$$
$$\leq \frac{A\ell(Q_i)}{\varepsilon^{n+1}}\|\nabla\psi\|_{L^\infty}\frac{A}{\ell(Q_i)} + \frac{A}{\varepsilon^{n+1}}\|\nabla\psi\|_{L^\infty}\|\eta_i\|_{L^\infty} \leq \frac{A}{\varepsilon^{n+1}},$$
$$|\nabla^2\varphi(\xi)| \leq \frac{A}{\varepsilon^{n+2}}\|\nabla^2\psi\|_{L^\infty}\|\eta_i\|_{L^\infty} + \frac{A}{\varepsilon^{n+1}}\|\nabla\psi\|_{L^\infty}\|\nabla\eta_i\|_{L^\infty}$$
$$+ \frac{A\ell(Q_i)}{\varepsilon^{n+1}}\|\nabla\psi\|_{L^\infty}\|\nabla^2\eta_i\|_{L^\infty} \leq \frac{A}{\ell(Q_i)\varepsilon^{n+1}}.$$

Finally we are in the position to use Lemma 3.2: $|\langle \varphi, \alpha_i \rangle| \leq \frac{A\,\ell(Q_i)\,\gamma(2Q_i \cap E)}{\varepsilon^{n+1}}$, or

$$(3.45) \qquad \|\psi_\varepsilon * \alpha_i\|_{L^\infty} \leq \frac{A\,\ell(Q_i)\,\gamma(2Q_i \cap E)}{\varepsilon^{n+1}}.$$

We plug (3.45) into (3.43) to obtain

$$(3.46) \qquad |U_\varepsilon^{\alpha_i}(z)| \leq \frac{A\,\ell(Q_i)\,\gamma(2Q_i \cap E)}{\varepsilon^n} \leq \frac{A\,\ell(Q_i)\,\gamma(2Q_i \cap E)}{\mathrm{dist}(z, 2Q_i)^n}$$

if $z \in \mathbb{R}^n \setminus 4Q_i$ and $\varepsilon > \frac{\mathrm{dist}(z, 2Q_i)}{2}$. The combination of (3.42) and (3.46) finishes the proof of Lemma 3.7. $\qquad\qquad\qquad\square$

Now, after the proof of this lemma, Theorem 3.6 is completely proved. It will serve as an important claim in the future: we will use it as an assumption of a certain Tb theorem which will prove the equivalence of γ and γ_+ in the case when $\gamma(E)$ is not much smaller than $\sum_{i=1}^{N} \gamma(2Q_i \cap E)$.

CHAPTER 4

From Distribution to Measure. Carleson Property

We fix a compact E. We recall that we have a fixed family $\{Q_i\}_{i=1}^N$ of cubes. We call them Whitney cubes because they will be constructed as Whitney cubes of a special neighborhood of E. These cubes have properties (3.13)-(3.15). There are smooth functions g_i with compact support in $2Q_i$. We are given a distribution S supported by E and such that

$$(4.1) \qquad \Phi * S \in L(E, 1) \,.$$

In (4.1), we used a special algorithm to built μ starting with S. Using g_i and μ, we built a complex measure ν. All of this has been done in the previous sections.

However, now we really need to modify the construction of measure ν.
Let

$$\Omega := \text{the interior of } \cup_{i=1}^N Q_i \,.$$

Let $\delta = \frac{1}{5} \operatorname{dist}(\partial\Omega, E)$. Then $\delta > 0$. Let us consider $\tau \in C_0^\infty(\mathbb{R}^n)$, with support in the unit ball centered at the origin, and let us assume that $\int \tau dx = 1$. Put $\tau_\delta := \frac{1}{\delta^n}\tau(\frac{\cdot}{\delta})$. Put $S_0 := \tau_\delta * S$. It is a smooth function with compact support E_0. Notice that E_0 lies in a δ-neighborhood of E, and in particular in $\cup_{i=1}^N Q_i$. In particular,

$$(4.2) \qquad \sum_{i=1}^N g_i = 1 \text{ on } E_0 \,.$$

From (4.1) one sees that

$$(4.3) \qquad \Phi * S_0 \in L(E_0, 1) \,,$$

just because $\Phi * S_0 = \tau_\delta * \Phi * S$, and the boundedness of the gradient by 1 is preserved by convolution with smooth function whose integral equals to 1.

So now instead of E we have E_0 and instead of distribution S we have a complex measure $S_0 dx$ (abusing the notation we will call this measure just S_0 sometimes). There are two advantages: 1) we will need to restrict $S_0 dx$ on closed cube $\bar{R}$, which is possible for measures but usually impossible for distributions, and 2) the next lemma allows us to estimate $S_0(\bar{R}) = \int_R S_0 dx$:

LEMMA 4.1. *Let s be a smooth function, and let $|\Psi * s dx|$ be a function bounded by 1. Let R be a cube. Then $|\int_R s dx| \le A\ell(R)^{n-1}$. If ν is a complex measure such that $|\Psi * f\nu|$ is a function bounded by 1 a.e. with respect to Lebesgue measure in $\mathbb{R}^n$, then $|\int_R d\nu| \le A\ell(R)^{n-1}$ and it does not matter whether we consider the closed or the open R. We can replace the cube R by the ball here. Only $\ell(R)$ should be replaced then by the radius of the ball.*

Proof. Just use Green's formula: $\int_R s\,dx = \int_R \Delta(\Phi * s)\,dx = \int_{\partial R} \frac{\partial \Phi * s}{\partial n}$. This and the bound on $|\Psi * s|$ finish the proof. The case of measure can be reduced to the s case if we mollify ν by considering $s = \psi_\varepsilon * \nu$, where ψ_ε is a smooth compactly supported function with support in $B(0, \varepsilon)$ and then tending ε to 0. $\square$

Let us return to Section 3.4. We use the same g_i's as before. We build exactly the same positive measure μ. But when we build ν we do not use the distribution S. We use the measure S_0.

We need to keep (3.24) valid. It followed from (3.19). So, because now S_0 replaces S, we need the analog of (3.19), namely we need, say

$$(4.4) \qquad |\langle S_0, g_i \rangle| \le 2A\gamma(2Q_i \cap E) \quad i = 1, \ldots, N,$$

where A is the constant from (3.19). But $|\langle S_0, g_i \rangle| = |\langle \tau_\delta * S, g_i \rangle|$, and $\tau_\delta * S$ converges weakly to distribution S. Therefore, (4.4) follows from (3.19) if δ is sufficiently small.

So let us fix δ as small as we have needed thus far. This defines S_0 and, consequently, ν.

Notice that all estimates we get so far are verbatim the same because of property (4.3). In particular, (3.31) of Theorem 3.6 is valid and (3.24) is valid (with may be slightly different constants). We do not use special a letter for this new measure ν; we prefer to keep the same letter. And we do not use special letters for constants in, say, (3.31) or (3.24); we keep the same letters.

But from now on ν is always the measure constructed by "old" μ and new S_0 instead of S exactly as this has been done in Section 3.4.

The next property can be called a Carleson property of measure ν (the one we have just built). Of course, in this case we deal with complex measure, and we are concerned with absolute values $|\nu(R)|$ rather than total variation of the measure on the cube. Recall that the classical Carleson property deals with positive measures.

THEOREM 4.2. *For any cube* $R \subset \mathbb{R}^n$

$$(4.5) \qquad |\nu(R)| \le A\,\ell(R)^{n-1}.$$

To prove the theorem we need the following lemmas.

For "small" cubes R this theorem follows from (3.24) and the following lemma.

LEMMA 4.3. *There exists a constant* A_0 *such that for any cube* $R \subset \mathbb{R}^n$ *and any Whitney cube* Q_i *with the property* $2Q_i \cap R \ne \emptyset$, *the relation* $\ell(Q_i) > \ell(R)/4$ *implies* $\mu(R) \le A_0\ell(R)^{n-1}$.

Proof. Notice that $R \subset 9\,Q_i$, and Whitney construction gives $\mathrm{card}\{j : Q_j \cap 9\,Q_i \ne \emptyset\} \le A_1$. Also $Q_j \asymp Q_i$ for those cubes. On each Q_j, just by construction of μ, $\mu(Q_j) \le A_2\,\ell(Q_j)^{n-1}$. Therefore,

$$\mu(R) \le \sum_{j:Q_j \cap 9\,Q_i \ne \emptyset} \mu(Q_j) \le A_2 \sum_{j:Q_j \cap 9\,Q_i \ne \emptyset} \ell(Q_j)^{n-1} \le A_1 A_2\,\ell(Q_i).$$

If

$$(4.6) \qquad \ell(Q_i) \le A_3\,\ell(R),$$

then we are done.

Otherwise, if the size of R is much smaller than the size of Q_i, the measure μ can come into R from at most one Whitney cube Q_j. But then there is nothing to prove; $\mu(R) \le A \ell(R)^{n-1}$ just because measure μ in every separate Q_j is very simple, just Lebesgue measure on a small sphere near the center of Q_j. $\square$

For other cubes R, one has $2Q_i \cap R \ne \emptyset \Rightarrow \ell(Q_i) \le \ell(R)/4$ for every cube Q_i from our family. For such cubes R we use

LEMMA 4.4. *There exists a constant A_1 such that for any cube $R \subset \mathbb{R}^n$ as above we have*

$$\text{(4.7)} \qquad \sum_{i\,:\,2Q_i\cap \partial R\ne \emptyset} |\langle g_i, S_0\rangle| \le A_1\, \ell(R)^{n-1}\,.$$

Proof. We have (4.3). Consequently by Lemma 3.1 we know that $|\langle g_i, S_0\rangle| \le A\,\ell(Q_i)^{n-1}$. Hence, $|\langle g_i, S_0\rangle| \le A\,\mathcal{H}^{n-1}(2Q_i \cap \partial R)$. But our Whitney cubes Q_i are such that $10Q_i$ have bounded (independently of E) multiplicity of intersection. So $\sum_{i\,:\,2Q_i\cap\partial R\ne\emptyset} |\langle g_i, S_0\rangle| \le A_1\,\ell(R)^{n-1}$. $\square$

Now we can prove Theorem 4.2.

Proof. We think that Q_i are closed cubes in this proof. Recall that $\mu|Q_i, \nu|Q_i$ are supported by a small sphere Σ_i contained in Q_i and centered at the center of Q_i. Let $I_R := \{i = 1, .., N : \Sigma_i \subset 2Q_i \subset \bar{R}\}$. Let $g_R := \sum_{i\in I_R} g_i$. We want to have

$$g_R(x) = 1 \; \forall x \in E_0 \cap \bar{R}\,.$$

Of course this might be false. Let x be a point of $E_0 \cap \bar{R}$ at which $g_R \ne 1$. On the other hand $\sum_{i=1}^N g_i(x) = 1$ as $x \in \operatorname{supp} S_0 = E_0$ (see (4.2)). Therefore, there exists $j \in [1, \ldots, N] \setminus I_R$ such that $x \in 2Q_j$ but $2Q_j$ is not a subset of R (otherwise j would be in I_R). Hence,

$$j \text{ is such that } 2Q_j \cap \partial R \ne \emptyset\,.$$

Let $L_R := \{j \notin I_R : 2Q_j \cap \partial R \ne \emptyset\}$. Then $G_R := g_R + \sum_{i\in L_R} g_i = 1$ on $E_0 \cap \bar{R}$. Moreover, this equality holds on a certain neighborhood of $E_0 \cap \bar{R}$. Let ψ_R denote the characteristic function of $\bar{R}$.

But $\psi_R - G_R = 0$ on $\operatorname{supp} S_0 \cap \bar{R}$. Slightly abusing the language, let us denote by S_R the restriction of S_0 on $\bar{R}$ (we use the fact that S_0 is a measure, not just a distribution). Then $\langle g_R, S_0\rangle = \langle g_R, S_R\rangle = \langle G_R, S_R\rangle - \sum_{i\in L_R}\langle g_i, S_R\rangle = \langle \psi_R, S_R\rangle - \sum_{i\in L_R}\langle g_i, S_R\rangle + \langle G_R - \psi_R, S_R\rangle =: I + II + III$. Clearly $III = 0$ as S_R is supported where $\psi_R - G_R = 0$. Let us see that $|I| \le A\ell(R)^{n-1}$. But this is obvious from Lemma 4.1 because we can write $I = \int_R S_0 dx$, and measure $S_0 dx$ satisfies the assumptions of this lemma.

Let us estimate $|II| \le A\ell(R)^{n-1}$. We want to use Lemma 4.4, but it involves S_0, not S_R. So instead we follow its proof. Let $2Q_j \cap \bar{R} \ne \emptyset$, and let x_j be the center of Q_j. Then

$$\langle g_j, S_R\rangle = \int_R g_j S_0 dx = \int_R g_j \cdot \Delta(\Phi * S_0 dx - (\Phi * S_0 dx)(x_j))$$

$$= \int_{R\cap 2Q_j} \Delta g_j(\Phi * S_0 dx - (\Phi * S_0 dx)(x_j))$$

$$+ \int_{\partial(R\cap 2Q_j)} \frac{\partial g_j}{\partial n}\cdot(\Phi * S_0 dx - (\Phi * S_0 dx)(x_j))$$

$$- \int_{\partial(R\cap 2Q_j)} g_j \cdot \frac{\partial(\Phi * S_0 dx)}{\partial n}\,.$$

The last integral is bounded by $A'\mathcal{H}^{n-1}(\partial(R \cap 2Q_j)) \leq A\mathcal{H}^{n-1}(\partial R \cap 2Q_j)$ because $\Phi * S_0 dx \in L(E, 1)$. The second integrand is bounded by $A |\nabla g_j| \ell(Q_j)$ for the same reason. Then the second integral is again bounded by $A'\mathcal{H}^{n-1}(\partial(R \cap 2Q_j)) \leq A\mathcal{H}^{n-1}(\partial R \cap 2Q_j)$. The first integrand is bounded by $A |\nabla^2 g_j| \ell(Q_j)$, and this gives again just the same estimate of the first integral. Finally we get $|II| \leq A\ell(R)^{n-1}$ because of finite multiplicity of intersection of $10Q_j$.

The estimates of I, II, III finally give us the estimate

$$(4.8) \qquad\qquad |\langle g_R, S_0 \rangle| \leq A \, \ell(R)^{n-1} \, .$$

To estimate $|\nu(R)|$ we should now remember that $\nu(R)$ is formed by summing up $\langle g_i, S_0 \rangle$ for $i \in I_R$ and adding $\langle g_i, S_0 \rangle \frac{\mu(R \cap \Sigma_i)}{\mu(\Sigma_i)}$ for some $i \in L_R$. Henceforth,

$$|\nu(R)| \leq |\langle g_R, S_0 \rangle| + \sum_{j \in L_R} |\langle g_j, S_0 \rangle| \, .$$

Estimates (4.7) and (4.8) finish the proof of (4.5). Theorem 4.2 is proved. $\square$

CHAPTER 5

Potential Neighborhood that has Properties
(3.13)–(3.14)

In this section we show the possibility of constructing cubes $\{Q_i\}_{i=1}^N$ satisfying properties (3.13)-(3.14). According to the idea of Tolsa [64] cubes Q_i are just Whitney cubes of the τ-neighborhood of E. It is very remarkable that τ is a small constant that does not depend on E! But this τ-neighborhood is not an Euclidean τ-neighborhood. The idea in [64] is to choose a special non-linear potential that peaks on E (is, say, equal to 1 on E). Then one takes a potential neighborhood of E. Namely, one considers an open set, where this potential is larger than a small absolute constant τ. To succeed in obtaining properties (3.13)-(3.14) for Whitney decomposition of this τ-neighborhood in the sense of potential, one needs to consider the potential of a special extremal measure. All of this happens in the classical potential theory, too. The difference is that to get (3.13)-(3.14), Tolsa needed to write a pertinent potential, which involved Melnikov–Menger's curvature. This is quite natural because the potential theory based on the Cauchy kernel is intimately related to this curvature. This has been discovered in [37] and [39]. But curvature has not been related to Riesz kernels for $n > 2$ thus far, so we have to modify Tolsa's idea.

However, we follow the same path, finding a certain extremal measure on E, building its special non-linear potential (without curvature), and proving that it almost peaks on E. Then we take an open set G, where this potential is larger than τ (τ depends on the dimension n only, but not on E). The Whitney decomposition of G will give us $\{Q_i\}_{i=1}^N$.

Recall that

$$(5.1) \qquad \gamma(E) \leq A \left(\operatorname{diam}(E)\right)^{n-1}.$$

We will do this under the auxiliary assumption

$$(5.2) \qquad \gamma_+(E) \leq \frac{1}{A_+ A}\left(\operatorname{diam}(E)\right)^{n-1}.$$

This is not at all an extra assumption, actually. The thing is that if it fails, then

$$\gamma_+(E) \leq \gamma(E) \leq A_+ A^2 \gamma_+(E).$$

And our main goal — the equivalence of two capacities — is fulfilled automatically. So everywhere below we assume (5.2).

The reader who is inclined to believe that cubes $\{Q_i\}_{i=1}^N$ satisfying properties (3.13)-(3.14) can be easily constructed can skip this section entirely and go immediately to Chapter 6. However, let us notice that this section differs from the same approximation construction in Tolsa's paper [64] for $n = 2$; moreover, the main tool used in [64] for this approximation construction, that is, Melnikov–Menger's curvature, is not available when $n > 2$. In the next section we start to study capacities with Calderón–Zygmund (CZ) kernels.

5.1. Capacities with Calderón–Zygmund (CZ) kernels

In this section we deal only with (positive, if otherwise not stated) measures satisfying the estimate from above:

$$(5.3) \qquad\qquad \sigma(B(x,r)) \leq r^m \,.$$

We call them *measures of order m*. No bound from below will be assumed. So we are always in a *nonhomogeneous* situation.

We wish to build the elements of potential theory. However, our kernel will not be positive. It will either be real valued but *alternating* (changing sign), or complex valued, or even vector valued. The elements of such Calderón–Zygmund potential theory appeared already in the recent literature: [63]-[65], [57].

In *nonhomogeneous* situations it is convenient to introduce the following

Definition. For a vector valued function f with values in $\mathbb{C}^d$, we define the following maximal function

$$M_\mu f(x) := \sup_{r>0} \frac{1}{\sigma(B(x,3r))} \int_{B(x,r)} |f|\, d\sigma \,.$$

Of course $|\cdot|$ is the Euclidean norm in $\mathbb{C}^d$.

For measures of order m one introduces *David's radius*.

Definition. Let $x \in \operatorname{supp} \sigma, r > 0$. Define

$$(5.4)$$
$$R = 3^J r, \ \ J := \max\{j \geq 0 : 2 \cdot 3^m \sigma(B(x, 3^{i-1}r)) < \sigma(B(x, 3^i r)), \ i = 1, \ldots, j\} \,.$$

The fact that $J = 0$ means that $\sigma(B(x, 3r)) \leq 2 \cdot 5^m \sigma(B(x, r))$; otherwise, we telescopically increase radii by multiplying it by 5 all the time until we get to $R = 3^J r$, which is the first such that

$$\sigma(B(x, 3R)) \leq 2 \cdot 3^m \sigma(B(x, R)) \,.$$

Our assumption that σ is of order m, and the fact that $x \in \operatorname{supp} \sigma$ guarantee us the existence of such a j. The existence of David's radius just says that a non-doubling measure cannot be non-doubling all the time (at least if the estimate from above of (5.3) holds).

We want to remind the reader of an important claim due to Guy David. Let T be a Calderón–Zygmund operator with kernel k of order m (see (8.7) and notice that m matches (5.3)). When we say that T is a Calderón–Zygmund operator, we include in the definition (as everywhere else in the literature) *the boundedness of T in $L^2(\mu)$*. To formulate the claim and the lemma that will follow it, we need

Notation. If $f \in L^2(\sigma)$, then $T^*_\sigma f(x) = \sup_{r>0} |T^r_\sigma f(x)|$, where

$$T^r_\sigma f(x) := \int_{y:|y-x| \geq r} k(x,y) f(y)\, d\sigma(y) \,.$$

If T acts in $L^2(\sigma)$, then there is no need to explain what $T_\sigma f$ is. Below we will need to understand $T_\sigma f$ also in some cases when we do not know a priori that T acts in $L^2(\sigma)$. In all such cases the Calderón–Zygmund kernel of T will be *antisymmetric* (unless otherwise stated): $k(x,y) = -k(y,x)$. The function f will be smooth always in such cases. And we understand $T_\sigma f$ in the sense of *canonical value operator*. Namely, if f, g are smooth test functions, then

$$\langle Tf, g \rangle = \frac{1}{2} \int \int k(x,y)[f(y)g(x) - f(x)g(y)] \, d\sigma(x)d\sigma(y) \,.$$

Claim (Guy David, [**19**]). Let T be a Calderón–Zygmund operator with kernel k of order m and let σ be a compactly supported measure that satisfies (5.3). Then

$$(5.5) \qquad (T_\sigma^* \mathbf{1})(x) \le D_1 \, M_\mu (T_\sigma \mathbf{1})(x) + D_2, \; \forall x \in \operatorname{supp} \sigma \,,$$

where D_1, D_2 are finite constants and D_1 depends on m and Calderón–Zygmund constants of k (see (8.7)), and D_2 depend on m and Calderón–Zygmund constants of k and on the norm $\|T\|$ of T in $L^2(\sigma)$.

Notice that saying that T is a Calderón–Zygmund operator we have already assumed that it is bounded in $L^2(\sigma)$. The lemma below says that the boundedness of T should not come into the estimate if the kernel k of T is antisymmetric.

LEMMA 5.1. *Let T be an operator (maybe not bounded) with CZ kernel k of order m. Let k be antisymmetric, and let σ be a compactly supported measure that satisfies (5.3). Then $\forall x \in \operatorname{supp} \sigma$*

$$(5.6) \qquad (T_\sigma^* \mathbf{1})(x) \le D_1 \, M_\mu (T_\sigma \mathbf{1})(x) + D_2,$$

where D_1, D_2 are finite constants and D_1, D_2 depend on m and Calderón–Zygmund constants of k.

Proof. The symbol $\langle \cdot \rangle_{G,\sigma}$ denotes the average over the set G with respect to measure σ. By C we denote various constants depending on Calderón–Zygmund constants of k, in particular, on m. Fix $x \in \operatorname{supp} \sigma, r > 0$, and let $R = R(x,r) = 3^J r$ be David's radius. We can write using (8.7), (5.3), and especially (5.4):

$$(5.7) \qquad |(T_\sigma^r) - (T_\sigma^{3R})f(x)| \le \sum_{j \le J} \frac{\sigma(B(x, 3^{j-1}r))}{(3^{j-1}r)^m} \le \frac{C \, \sigma(B(x, 3R))}{R^m} \le C \,.$$

Denote $\sigma_{3R} := \sigma | \mathbb{R}^n \setminus B(x, 3R)$. Now compare $T_\sigma^{3R} \mathbf{1}(x)$ and $\langle T_{\sigma_{3R}} \mathbf{1}(z) \rangle_{B(x,LR),\sigma}$, where the number $L \in [1, 2]$ will be chosen appropriately:

$$T_\sigma^{3R} \mathbf{1}(x) - \langle T_{\sigma_{3R}} \mathbf{1}(z) \rangle_{B(x,LR),\sigma}$$

$$= \int_{y:|y-x| \ge 3R} d\sigma(y) \int k(z,y) [d\delta_{\{x\}} - \frac{d\sigma | B(x,LR)}{\sigma(B(x,LR))}] \,.$$

The measure in $[\cdot]$ has zero integral and mass 2. It is supported by $\frac{1}{2}B(x, 3R)$. Hence, by the second Calderón–Zygmund property of the kernel,

$$|(T_\sigma^{3R}) \mathbf{1}(x) - \langle (T_{\sigma_{3R}}) \mathbf{1}(z) \rangle_{B(x,LR),\sigma}|$$

$$\le \int_{y:|y-x| \ge 3R} d\sigma(y) \int |k(z,y) - k(x,y)| |d\delta_{\{x\}} - \frac{d\sigma | B(x,LR)}{\sigma(B(x,LR))}|$$

$$\le C \, R^\varepsilon \int_{y:|y-x| \ge 3R} \frac{1}{|x-y|^{m+\tau}} d\sigma(y) \le C \,.$$

We used (5.3) in the last inequality. Now, of course,

$$\langle T_{\sigma_{3R}} \mathbf{1}(z) \rangle_{B(x,LR),\sigma} = \langle T_\sigma \mathbf{1}(z) \rangle_{B(x,LR),\sigma} - \langle T_{\sigma | B(x,3R)} \mathbf{1}(z) \rangle_{B(x,LR),\sigma} =: I - II \,.$$

Term I is easy to estimate:

$$|I| \le \frac{\sigma(B(x,3R))}{\sigma(B(x,LR))}(M_\mu T_\sigma \mathbf{1})(x) \le \frac{\sigma(B(x,3R))}{\sigma(B(x,R))}(M_\mu T_\sigma \mathbf{1})(x) \le 2 \cdot 3^m (M_\mu T_\sigma \mathbf{1})(x).$$

This is by the definition of David's radius R.

We are left to estimate $|II| = |T_{\sigma|B(x,3R)}\mathbf{1}(z)\rangle_{B(x,LR),\sigma}|$. This will be done by the choice of L. Let $G := B(x,3R) \setminus B(x,LR)$, $F := B(x,LR)$.

$$T_{\sigma|B(x,3R)}\mathbf{1}(z)\rangle_{B(x,LR),\sigma} = \frac{1}{\sigma(B(x,LR))}\langle T_\sigma \mathbf{1}_{G\cup F}, \mathbf{1}_F\rangle = \frac{1}{\sigma(B(x,LR))}\langle T_\sigma \mathbf{1}_G, \mathbf{1}_F\rangle.$$

We used here the antisymmetry of k.

Now we want to prove that there exists $L \in [1,2]$ such that

(5.8)
$$\sigma\{z \in \mathbb{R}^n : (1-\tau)L\,R \le |z - x| \le (1+\tau)L\,R\} \le C\,\tau\,\sigma(B(x,3R)) \ \forall \tau \in [0,1].$$

In fact, let ρ be a measure on $B(0,5)$ given by

$$\rho(E) := \sigma(R\,E).$$

In particular, the fact that σ is of order m implies that $\|\rho\| \le R^m$. Project ρ radially on $[0,5]$ to get measure κ. Put

$$M\kappa(x) := \sup_{\tau>0} \frac{\kappa\{(x-\tau, x+\tau)\}}{\tau}.$$

One has a weak type estimate

$$|\{x : M\kappa(x) > 1000\|\kappa\|\} \le A\frac{\|\kappa\|}{1000\|\kappa\|} \le \frac{A}{1000},$$

with an absolute constant less than 500. In particular, for at least one half of L's in $[1,2]$ (in the sense of Lebesgue measure) we have

$$M\kappa(L) \le 1000\|\kappa\| = 1000\|\rho\|.$$

Choose such an L. Then $\kappa\{(L-\tau, L+\tau)\} \le 1000\|\rho\|\tau$, or $\kappa\{(L(1-\tau), L(1+\tau))\} \le 2000\|\rho\|\tau = 2000\tau\sigma(B(x,3R))$. This proves (5.8).

Now we continue the estimate of II. To this end we want to estimate $\langle \mathbf{1}_G, \mathbf{1}_F\rangle$, where $F = B(x,LR), G = B(x,3R) \setminus B(x,LR)$. Let $y \in G, d = d(y) := \mathrm{dist}(y,F)$. Then

(5.9)
$$|T_\sigma \mathbf{1}_F(y)| \le C \log \frac{3R}{d}.$$

In fact,

$$|T_\sigma \mathbf{1}_F(y)| \le c \sum_{i=0}^{[\log R/d]+1} \frac{\sigma(B(y,2^{i+1}d))}{(2^i\,d)^m} \le C \sum_{i=0}^{[\log R/d]+1} 2^m \le C \log \frac{3R}{d}.$$

Now we get, combining (5.9) and (5.8),

$$|\langle \mathbf{1}_G, \mathbf{1}_F\rangle| \le \int_G |T_\sigma \mathbf{1}_F(y)|\,d\sigma(y) \le C \int_G \log \frac{3R}{d(y)}\,d\sigma(y)$$

$$\le C\,\sigma(B(x,3R)) \int_0^{3R} \log \frac{3R}{t}\,\frac{dt}{R}.$$

Hence

$$|\langle \mathbf{1}_G, \mathbf{1}_F\rangle| \le C\,\sigma(B(x,3R)).$$

So

$$|II| \leq C \, \frac{\sigma(B(x,3R))}{\sigma(B(x,LR))} \leq C \, \frac{\sigma(B(x,3R))}{\sigma(B(x,R))} \leq C \, 2 \cdot 3^m \, ,$$

because R is David's radius. Lemma 5.1 is finally proved. $\qquad\square$

We are going to introduce several capacities with *alternating* kernel (actually vector valued kernel). It will be important to work with all of them simultaneously. Let k be an *antisymmetric* Calderón–Zygmund kernel, $k : \mathbb{R}^n \times \mathbb{R}^n \to \mathbb{C}^d$. Let it be of order m (see (8.7)). Let E be a compact in $\mathbb{R}^n$. The symbol $M_+(E)$ denotes the set of nonnegative measures on E.

Notation. $\Sigma := \{\sigma \in M_+(E) : \sigma(B(x,r)) \leq r^m \; \forall x \in \mathbb{R}^n, \forall r > 0\}.$

Definitions.

$$\gamma_{op}(E) := \sup\{\|\sigma\| : \sigma \in \Sigma, \|T : L^2(\sigma) \to L^2(\sigma)\| \leq 1\} \, ,$$

$$\gamma_+^*(E) := \sup\{\|\sigma\| : \sigma \in \Sigma, \; T_\sigma^* \mathbf{1} \leq 1 \text{ on } E\} \, ,$$

$$\gamma_{T1}(E) := \sup\left\{\|\sigma\| : \sigma \in \Sigma, \int_E |T_\sigma \mathbf{1}|^2 \, d\sigma \leq \|\sigma\|\right\} \, ,$$

Let us consider "cut-off" kernels. Fix a smooth function ψ having compact support. And consider $k_\tau(x,y) := \psi(\frac{|x-y|}{\tau}) k(x,y)$. The corresponding operator with *bounded antisymmetric* kernel k_τ is called $T^{(\tau)}$.

We will need $\gamma_{T^{(\tau)}1}$.

LEMMA 5.2. $\limsup_{\tau \to 0} \gamma_{T^{(\tau)}1} \leq A \, \gamma_{T1}$.

THEOREM 5.3. $\gamma_{T1} \asymp \gamma_{op}$.

THEOREM 5.4. $A^{-1} \gamma_{T1} \leq \liminf_{\tau \to 0} \gamma_{T^{(\tau)}1} \leq \limsup_{\tau \to 0} \gamma_{T^{(\tau)}1} \leq A \, \gamma_{T1}$.

These results require proofs, which will be given now. But first notice that there is one trivial comparison:

$$\tag{5.10} \gamma_+^* \leq \gamma_{T1} \, .$$

Indeed, if $\sigma \in \Sigma$ is such that $T_\sigma^* \mathbf{1} \leq 1$ on E, then it is easy to see that $\int_E |T_\sigma \mathbf{1}|^2 \, d\sigma \leq \|\sigma\|$, and this means that any σ participating in γ_+^* participates in γ_{T1} too. So (5.10) is proved.

Remark. Theorem 5.3, (5.10) and (5.11) show that all capacities are equivalent.

Now we are going to prove Theorem 5.3.

Proof. We start with proving

$$\tag{5.11} \gamma_{op}(E) \leq C \, \gamma_+^*(E) \, .$$

Let σ participate in $\gamma_{op}(E)$, i.e., $\sigma \in \Sigma$, and $\|T : L^2(\sigma) \to L^2(\sigma)\| \leq 1$. The reader can find in [**44**] or in Section 5.3 below that then there exists $h \in L^\infty(\sigma), 0 \leq h \leq 1$, such that

$$\tag{5.12} \int_E h \, d\sigma \geq \frac{1}{2} \|\sigma\| \, ,$$

$$\tag{5.13} \|T_\sigma h\|_{L^\infty(\sigma)} \leq C \, .$$

Here C depends only on n, m, and Calderón–Zygmund parameters of the kernel of T. (It depends also on the norm of T in $L^2(\sigma)$, but this norm is assumed to be bounded by 1.)

Use again our knowledge of nonhomogeneous L^p theory, namely, the lemma of Cotlar (our kernel is of order m and the measure σ is of order m, so we can apply Cotlar's lemma from Section 5.3 below):

$$(5.14) \qquad T_\sigma^* h(x) \le C_1 (M_\mu |T_\sigma h|^\beta)^{1/\beta}(x) + C_2 M_\mu h(x) \le C_3 \,,$$

where $\beta \in (0,1)$ is arbitrary fixed number, and C_1, C_2 depend only on n, m, and Calderón–Zygmund parameters of the kernel of T, on β (which we fix to be $1/2$), and on $\|T\|$ in $L^2(\sigma)$, but this norm is assumed to be bounded by 1. This is why C_3 in the last inequality depends only on n, m, and Calderón–Zygmund parameters of the kernel of T. Consider $C_4 := \max(1, C_3), \sigma_1 := h\, d\sigma / C_4$. Then $\sigma_1 \in \Sigma$, and (5.14) can be rewritten as

$$T_{\sigma_1}^* \mathbf{1} \le 1 \,,$$

and so, σ_1 participates in $\gamma_+^*(E)$. Hence,

$$\gamma_+^*(E) \ge \|\sigma_1\| \ge C_4^{-1} \int h\, d\sigma \ge (2C_4)^{-1} \|\sigma\| \,,$$

by (5.12). Here σ is any measure participating in γ_{op}. Henceforth we proved (5.11).

We are left to show that (5.15) holds:

$$(5.15) \qquad \gamma_{T1}(E) \le C\, \gamma_{op}(E) \,.$$

We will be using Calderón–Zygmund operators with suppressed kernels. So the reader is referred to first read Section 8.1.

Let $\sigma \in \Sigma$ participate in γ_{T1}:

$$\int |T_\sigma \mathbf{1}|^2 \, d\sigma \le \|\sigma\| \,.$$

We are not given the boundedness of T (unlike the case when $\sigma \in \Sigma$ participates in γ_{op}). So we cannot use Cotlar's lemma from Section 5.3. But we prepared Lemma 5.6 for this occasion:

$$T_\sigma^* \mathbf{1} \le C_1 M_\mu T_\sigma \mathbf{1} + C_2 \,.$$

Here C_1, C_2 depend on m, n, and Calderón–Zygmund constants of the kernel. Combining the last two inequalities we see that

$$\int |T_\sigma^* \mathbf{1}|^2 \, d\sigma \le C_0 \|\sigma\| \,.$$

Then

$$(5.16) \qquad \text{for the set } F := \{x \in E : T_\sigma^* \mathbf{1}(x) \le \sqrt{2C_0}\} \text{ we have } \sigma(F) \ge \frac{1}{2}\|\sigma\| \,.$$

Our goal now is to deduce from (5.16) that

$$(5.17) \qquad \|T_\sigma : L^2(F, d\sigma) \to L^2(F, d\sigma)\| \le c' = C(m, n, CZ) \,.$$

As soon as this is done (5.15) is proven: in fact, then $\frac{1}{C}\sigma|F$ ($C := \max(1, c')$) participates in $\gamma_{op}(E)$, and so

$$\gamma_{op}(E) \ge C^{-1}\sigma(F) \ge (2C)^{-1}\|\sigma\| \,.$$

But σ was an arbitrary measure participating in $\gamma_{T1}(E)$. Thus $\gamma_{op}(E) \geq (2C)^{-1}\gamma_{T1}(E)$, which is (5.15).

To prove (5.17) let us consider the open set $G := D \setminus F$, where D is a large open ball containing E deep inside. For any $z \in G \cap E = E \setminus F$, set $\varepsilon(z) = \mathrm{dist}(z, F)$. First of all estimate for any $\varepsilon \geq \varepsilon(z)$

$$T_\sigma^\varepsilon \mathbf{1}(z) := \int_{y:|y-z|\geq\varepsilon} k(z,y)\,d\sigma(y) = \int_{y:|y-z|\geq 2\varepsilon} \cdots + \int_{y:\varepsilon\leq|y-z|<2\varepsilon} \cdots =: I + II\,.$$

Obviously, $|II| \leq C$ as $\sigma(B(z,2\varepsilon)) \leq 2^m\varepsilon^m$ ($\sigma \in \Sigma$ is used) and $|k(z,y)| \leq \frac{C}{|z-y|^m} \leq \frac{C}{\varepsilon^m}$ in the second integrand.

Let x_0 be a closest point to z on F. We write

$$I = \int_{y:|y-z|\geq 2\varepsilon} k(x_0,y)\,d\sigma(y) + \int_{y:|y-z|\geq 2\varepsilon} [k(z,y) - k(x_0,y)]\,d\sigma(y) =: III + IV\,.$$

Now $|III| \leq \sqrt{2C_0}$ by the fact that $x_0 \in F$; see (5.16). On the other hand, by the second property (8.7) $|k(z,y) - k(x_0,y)| \leq \frac{C\varepsilon}{|z-y|^{m+1}}$ in the integrand of IV. Combining this with $\sigma \in \Sigma$, we get $|IV| \leq C$. We proved (5.18):

$$(5.18) \qquad |T_\sigma^\varepsilon \mathbf{1}(z)| \leq C\ \forall \varepsilon \geq \varepsilon(z)\,.$$

Now the reader is referred to first read Section 8.1 for the notion of the suppressed operator. We will suppress k by the function $\Theta(z) := \mathrm{dist}(z, F)$. We consider k_Θ, T^Θ (see (8.1) and the definition of T^Θ below it). Our next step in proving (5.17) is to show

$$(5.19) \qquad |T_\sigma^{\Theta,\varepsilon} \mathbf{1}(z) - T_\sigma^\varepsilon \mathbf{1}(z)| \leq C\ \forall \varepsilon \geq \varepsilon(z)\,.$$

Indeed, $(\varepsilon(z) = \Theta(z))$

$$I := |T_\sigma^{\Theta,\varepsilon} \mathbf{1}(z) - T_\sigma^\varepsilon \mathbf{1}(z)| \leq \int_{y:|y-z|\geq\Theta(z)} |k^\Theta(z,y) - k(z,y)|\,d\sigma(y)\,.$$

But by definition $|k^\Theta(z,y) - k(z,y)| \leq |k(z,y)|\frac{|k(z,y)|^2\Theta(z)^m\Theta(y)^m}{1+|k(z,y)|^2\Theta(z)^m\Theta(y)^m}$. In our case $\Theta(z) \leq |y-z|$, $\Theta(y) \leq \Theta(z) + |y-z| \leq 2|y-z|$. Hence, in the range of y's we need, we have

$$|k^\Theta(z,y) - k(z,y)| \leq |k(z,y)|\frac{\Theta(z)^m\Theta(y)^m}{|k(z,y)|^{-2}} \leq \frac{C\,2^m\,\Theta(z)^m}{|k(z,y)|^{-2}} \leq \frac{C\,\Theta(z)^m}{|z-y|^{2m}}\,.$$

Therefore, using $\sigma \in \Sigma$ again, we get

$$I \leq \int_{y:|y-z|\geq\Theta(z)} \frac{\Theta(z)^m}{|z-y|^{2m}}\,d\sigma(y) \leq C\sum_{k\geq 0} \frac{(2^{k+1}\Theta(z))^m\Theta(z)^m}{(2^k\Theta(z))^{2m}} \leq C\,.$$

We proved (5.19). Combine it with (5.18) to get

$$|T_\sigma^{\Theta,\varepsilon} \mathbf{1}(z)| \leq C\ \forall \varepsilon \geq \varepsilon(z)\,.$$

We want $T_\sigma^{\Theta,*} \mathbf{1}(z) \leq C$. In view of the previous inequality we need only to see that for $\eta \in (0, \varepsilon(z))$,

$$\left|\int_{y:|y-z|\leq\eta} k_\Theta(z,y)\,d\sigma(y)\right| \leq \frac{C}{\Theta(z)^m}\sigma(B(z,\Theta(z))) \leq C\,.$$

We used that $\eta \leq \varepsilon(z) = \Theta(z)$ and that $\sigma \in \Sigma$. We also used the property of suppressed kernels. Finally we get

$$T_\sigma^{\Theta,*}\mathbf{1}(z) = \sup_{\eta > 0} |T_\sigma^{\Theta,\eta}\mathbf{1}|(z) \leq C \ \forall z \in E \setminus F \,.$$

However this is true on F, too. In fact, if $x \in F$, then $k_\Theta(x,y) = k(x,y)$ for all y, and so $T_\sigma^{\Theta,*}\mathbf{1}(x) = T_\sigma^*\mathbf{1}(x) \leq \sqrt{2C_0}$ by (5.16). Thus,

$$(5.20) \qquad\qquad T_\sigma^{\Theta,*}\mathbf{1}(x) \leq C \ \forall x \in E \,.$$

The general theory of antisymmetric Calderón–Zygmund operators with kernels of order m with respect to measure of order m shows (see Section 5.3, it is a so-called nonhomogeneous $T1$ theorem) that

$$(5.21) \qquad \|T_\sigma^\Theta : L^2(E,d\sigma) \to L^2(E,d\sigma)\| \leq C = C(n,m,CZ) \,.$$

But for $f,g \in L^2(E,d\sigma)$, which are supported on F, one has (see the definition of kernel k_Θ and use the fact that $\Theta = 0$ on F)

$$\langle T_\sigma f, g \rangle = \langle T_\sigma^\Theta f, g \rangle \,.$$

This and (5.21) implies (5.17). This proves (5.15) as we have already learned. Thus, Theorem 5.3 is completely proved. $\qquad\square$

Proof of Lemma 5.2. Let us recall that

$$T^\varepsilon f(x) = T_\sigma^\varepsilon f(x) := \int_{y:|y-x|\geq\varepsilon} k(x,y)f(y)\,d\sigma(y) \,.$$

$$T^{(\varepsilon)} f(x) = T_\sigma^{(\varepsilon)} f(x) := \int_{y:|y-x|\geq\varepsilon} k(x,y)\psi\Big(\frac{|x-y|}{\varepsilon}\Big)f(y)\,d\sigma(y) \,,$$

where $\psi \in C^\infty(\mathbb{R}^n)$, equals 1 on $\mathbb{R}^n \setminus B(0,1)$, and equals 0 on $B(0,1/2)$. The first operator is not a Calderón–Zygmund operator; the second one is a Calderón–Zygmund operator. Also

$$|T^\varepsilon f(x) - T^{(\varepsilon)} f(x)| \leq C \sup_{r>0} \frac{\int_{B(x,r)} |f(y)|\,d\sigma(y)}{r^m} \,.$$

In particular, as $\sigma \in \Sigma$, we have

$$(5.22) \qquad\qquad |T^\varepsilon \mathbf{1}(x) - T^{(\varepsilon)}\mathbf{1}(x)| \leq C \,.$$

Now to prove that $\limsup_{\tau\to 0} \gamma_{T^{(\tau)}\mathbf{1}}(E) \leq C\,\gamma_{T\mathbf{1}}(E)$, we, first of all, replace all these γ's by $\gamma_+^*(E)$ (with respect to Calderón–Zygmund operators $T^{(\tau)}, T$). This is possible by the previous Theorem 5.3. So let $\tau_n \to 0$ and *decrease monotonically*. Let $\sigma_n \in \Sigma$, $\|\sigma_n\| = \gamma_{T^{(\tau_n)},+}^*(E)$ and

$$T_{\sigma_n}^{(\tau_n),*}\mathbf{1} \leq 2 \text{ on } E \,.$$

Without loss of generality we think that $\sigma_n \to \sigma_0$ weakly. Let us fix $\tau \in [\tau_{m+1}, \tau_m]$, $n > m$. Then by definition of what $*$ is and by (5.22), we get

$$T_{\sigma_n}^{(\tau)}\mathbf{1} \leq C + 2 \text{ on } E \,.$$

So

$$T_{\sigma_0}^{(\tau)}\mathbf{1} \leq C + 2 \text{ on } E \ \forall \tau > 0 \,.$$

Using (5.22) again, we get that for σ_0 (and, by the way, $\sigma_0 \in \Sigma$)

$$T_{\sigma_0}^*\mathbf{1} \leq 2C + 2 \,.$$

Hence, measure $\frac{1}{2C+2}\sigma_0$ participates in $\gamma_{T,+}^*(E)$. Henceforth,

$$\gamma_{T,+}^*(E) \geq \frac{1}{2C+2} \lim_n \|\sigma_n\| = \frac{1}{2C+2} \limsup_{\tau \to 0} \gamma_{T^{(\tau)}\mathbf{1}}(E)$$

and Lemma 5.2 is proved. $\qquad\qquad\qquad\qquad\qquad\qquad\qquad\qquad\qquad\square$

Remark. This lemma means the following: "cut-off" operators $T^{(\tau)}$ are not so singular as T; it is easier to work with them. But precisely because of that, it is quite imaginable that capacities built for them are much larger than those built for T. Just restrictions on measures imposed by $T^{(\tau)}$ could have been much more relaxed than restrictions imposed by T. If τ is far from 0 this is the case. But the lemma says that for small τ, we have the same (equivalent) capacities. This is very convenient because for small τ, operator $T^{(\tau)}$ is much easier to work with than T.

Proof of Theorem 5.4. After the previous lemma we are left to prove a simple inequality:

$$\gamma_{T\mathbf{1}}(E) \leq C \liminf_{\tau \to 0} \gamma_{T^{(\tau)}\mathbf{1}}(E).$$

Again, due to Theorem 5.3, each of the quantities can be replaced by $\gamma_+^*(E)$ (with respect to $T, T^{(\tau)}$ correspondingly). Suppose $\sigma \in \Sigma$ participates in $\gamma_{T,+}^*(E)$; that is,

$$T_\sigma^*\mathbf{1} \leq 1 \text{ on } E.$$

Then it is easy to see that (use (5.22))

$$T_\sigma^{(\tau),*}\mathbf{1} \leq C \text{ on } E.$$

So, $\gamma_{T^{(\tau)},+}^*(E) \geq \frac{1}{C}\|\sigma\|$ for every $\sigma \in \Sigma$ participating in $\gamma_{T,+}^*(E)$. So

$$\inf_{\tau > 0} \gamma_{T^{(\tau)},+}^*(E) \geq \frac{1}{C}\gamma_{T,+}^*(E).$$

The proof of Theorem 5.4 is completed.

5.2. Variational capacity and extremal measures

We start with repeating the main inequalities of the previous section:

$$\limsup_{\tau \to 0} \gamma_{T^{(\tau)}\mathbf{1}}(E) \leq C_1 \gamma_{T\mathbf{1}}(E),$$

$$\gamma_{T\mathbf{1}}(E) \leq C_2 \gamma_{op}.$$

Choose τ_0 so small that

(5.23) $$\gamma_{T^{(\tau_0)}\mathbf{1}}(E) \leq 2C_1 C_2 \gamma_{op}(E).$$

Definition. $S := T^{(\tau_0)}$. This is a Calderón–Zygmund operator of order m. Of course S, τ_0 depend on E.

Definition. Let us introduce the variational capacity

$$\gamma_{var}(E) := \sup_{\sigma \in \Sigma} \frac{\|\sigma\|^2}{\|\sigma\| + \int_E |S_\sigma\mathbf{1}|^2 \, d\sigma}.$$

THEOREM 5.5. *The supremum is attained and*

(5.24) $$\gamma_{var}(E) \asymp \gamma_{op}(E),$$

where the constants of comparison depend on n, m, and Calderón–Zygmund constants of the kernel, but do not depend on E.

Proof. Consider

$$F(\sigma) := \frac{\|\sigma\|^2}{\|\sigma\| + \int_E |S_\sigma \mathbf{1}|^2 \, d\sigma}, \quad \sigma \in \Sigma(E).$$

Notice that Σ is a compact subset of the set of positive measures on compact E. So if $\sigma_n \to \sigma_0$ weakly, then $\|\sigma_n\| \to \|\sigma_0\|$. Also,

$$\int_E |S_{\sigma_n} \mathbf{1}|^2 \, d\sigma_n \to \int_E |S_{\sigma_0} \mathbf{1}|^2 \, d\sigma_0.$$

This is because the kernel of $S = T^{(\tau_0)}$ is continuous. Continuous functional F attains its maximum on compact $\Sigma(E)$.

We are left to prove (5.24). By Theorems 5.4, 5.3, it is enough to compare $\gamma_{var}(E)$ with $\gamma_{S\mathbf{1}}(E)$. Let us start by proving that any extremal (for the functional F) σ_0 satisfies

$$(5.25) \qquad\qquad \int_E |S_{\sigma_0} \mathbf{1}|^2 \, d\sigma_0 \le \|\sigma_0\|.$$

If not, then $\int_E |S_{\sigma_0} \mathbf{1}|^2 \, d\sigma_0 = N \|\sigma_0\|$ with $N > 1$. Consider $\sigma_1 := \frac{1}{N^{1/2}} \sigma_0$. It belongs to Σ and

$$F(\sigma_1) = \frac{\sigma_0(E)^2 \, N^{-1}}{\sigma_0(E) \, N^{-1/2} + N^{-1/2} \, \sigma_0(E)} = \frac{1}{2N^{1/2}} \, \sigma_0(E) > \frac{1}{1+N} \, \sigma_0(E) = F(\sigma_0)$$

and we come to the contradiction with the assumption that σ_0 is a maximizer.

Given (5.25) we get by definition

$$\gamma_{var}(E) = F(\sigma_0) < \|\sigma_0\| \le \gamma_{S\mathbf{1}}(E).$$

Conversely,

$$\frac{1}{2} \gamma_{S\mathbf{1}}(E) \le \gamma_{var}(E).$$

In fact, for any σ participating in $\gamma_{S\mathbf{1}}(E)$, one has $\int_E |S_\sigma \mathbf{1}|^2 \, d\sigma \le \|\sigma\|$. Hence (as $\sigma \in \Sigma$ also)

$$\frac{\|\sigma\|}{2} \le \frac{\|\sigma\|^2}{\|\sigma\| + \int_E |S_\sigma \mathbf{1}|^2 \, d\sigma} \le \gamma_{var}(E)$$

and we are done. Theorem 5.5 is completely proved. $\square$

Definition. From here to the end of the section we keep the name σ_0 for a maximizer for the functional F. We recall that the kernel of S is vector valued. We use

$$S_{\sigma_0} S_{\sigma_0} \mathbf{1} := \sum_{i=1}^{d} \int \int k_i(x, y) \overline{k_i(y, z)} \, d\sigma_0(z) d\sigma_0(y).$$

The next is an important *variational lemma*.

LEMMA 5.6. *Suppose that $H \in M_+(E)$ is a positive measure on E such that*

$$\sigma_\lambda := \sigma_0 + \lambda H \in \Sigma(E) \text{ for all } \lambda \in [0, \lambda_0], \lambda_0 > 0.$$

Then

$$(5.26) \qquad \frac{\|\sigma_0\| + 2 \int_E |S_{\sigma_0} \mathbf{1}|^2 \, d\sigma_0}{\|\sigma_0\|} \le \int \left[|S_{\sigma_0} \mathbf{1}|^2 - 2\Re \, S_{\sigma_0} S_{\sigma_0} \mathbf{1} \right] \frac{dH}{H(E)}.$$

Proof. Notice that

(5.27)
$$\int S_H \mathbf{1} \cdot S_{\sigma_0} \mathbf{1} \, d\sigma_0 = -\int S_{\sigma_0} S_{\sigma_0} \mathbf{1} \, dH \,.$$

Inequality $F(\sigma_0) \geq F(\sigma_\lambda)$ can be rewritten as follows:

$$\lambda\Big[\sigma_0(E)^2 (H(E) + \int |S_{\sigma_0}\mathbf{1}|^2 \, dH - 2\Re \int S_{\sigma_0}S_{\sigma_0}\mathbf{1}\,dH)$$
$$- 2\,\sigma_0(E)H(E)(\sigma_0(E) + \int |S_{\sigma_0}\mathbf{1}|^2\,d\sigma_0)\Big]$$
$$+ \lambda^2 \, \text{Term} + \lambda^3 \, \sigma_0(E)^2 \int |S_H \mathbf{1}|^2 \, dH \geq 0\,.$$

Kernel S is continuous, so, of course, Term is finite as well as $\int |S_H \mathbf{1}|^2 \, dH$. Therefore, dividing by λ and making $\lambda \to 0$, one gets

$$\sigma_0(E)(H(E) + \int |S_{\sigma_0}\mathbf{1}|^2 \, dH - 2\Re \int S_{\sigma_0}S_{\sigma_0}\mathbf{1}\,dH)$$
$$- 2\,H(E)(\sigma_0(E) + \int |S_{\sigma_0}\mathbf{1}|^2\,d\sigma_0) \geq 0\,.$$

Divide by $\sigma_0(E)H(E)$ to get

$$1 + \int |S_{\sigma_0}\mathbf{1}|^2 \, \frac{dH}{H(E)} - 2\Re \int S_{\sigma_0}S_{\sigma_0}\mathbf{1} \, \frac{dH}{H(E)} \geq 2\frac{\sigma_0(E) + \int_E |S_{\sigma_0}\mathbf{1}|^2\,d\sigma_0}{\sigma_0(E)}\,,$$

which is exactly (5.26). $\qquad\square$

Definition. Let $M\sigma(x) := \sup_{r>0} \frac{\sigma(B(x,r))}{r^m}$. Recall that S has a vector valued kernel, and the ith entry of this kernel is a complex or real valued Calderón–Zygmund kernel which is continuous. The corresponding integral operator is denoted by S^i, $i = 1, \ldots, d$. We want to use from now on the notation (f is a scalar valued function)

$$S^*_\sigma f := S^{1,*}_\sigma f + \cdots + S^{d,*}_\sigma f\,.$$
$$S^*_\sigma S_{\sigma_0} f := S^{1,*}_\sigma (\overline{S^1_{\sigma_0} f}) + \cdots + S^{d,*}_\sigma (\overline{S^d_{\sigma_0} f})\,.$$

(The fact that one σ is without subscript 0 is not a misprint.) Consider the following potential

$$U^{\sigma_0}(x) := M\sigma_0(x) + S^*_{\sigma_0}\mathbf{1}(x) + S^*_\sigma S_{\sigma_0}\mathbf{1}\,.$$

THEOREM 5.7. *Let E be a finite union of cubes or a finite union of k dimensional smooth manifolds, $m \leq k < n$. Choose τ_0 and corresponding S. Let σ_0 be a maximizer of the variational problem. There exists $\alpha_0 = \alpha_0(n, m, CZ) > 0$ such that*

(5.28)
$$U^{\sigma_0} \geq \alpha_0 \quad \text{everywhere on } E\,.$$

Proof. Choose $\alpha_1 \in (0, 80^{-m})$. If $x_0 \in E$ is such that $M\sigma_0(x_0) \geq \alpha_1$, then we are done at x_0. So let

(5.29)
$$M\sigma_0(x_0) < \alpha_1\,.$$

Recall that the kernel S was defined with the help of a certain small parameter τ_0. Choose $\varepsilon_1 \leq \tau_0$, $R_0 := \frac{\varepsilon_1}{10}$, $B_0 := B(x_0, R_0)$. Set $\sigma_{00} := \sigma_0|2\,B_0$. Put

$$G := \{y \in B_0 \cap E : M\sigma_0(y) \leq \frac{1}{4^m}\}\,.$$

We want to prove that the complement of G in $B_0 \cap E$ is small. If $y \in (B_0 \cap E) \setminus G$, then

$$(5.30) \qquad \exists r = r(y) > 0 \text{ such that } \frac{\sigma_0(B(y,r))}{r^m} > \frac{1}{4^m}.$$

If in addition $r > \frac{\varepsilon_1}{20} = \frac{R_0}{2}$, then

$$\frac{1}{(4 \cdot 20)^m} \leq \frac{\sigma_0(B(y,r))}{20^m\, r^m} \leq \frac{\sigma_0(B(x_0, 20r))}{(20r)^m} < \alpha_1,$$

by (5.29). This shows that r in (5.30) is at most $\frac{\varepsilon_1}{20}$. But then

$$\sigma_{00}(B(y,r)) = \sigma_0(B(y,r)).$$

This and (5.30) mean that

$$(5.31) \qquad \forall y \in (B_0 \cap E) \setminus G \,\exists r = r(y) > 0 \text{ such that } \frac{\sigma_{00}(B(y,r))}{r^m} > \frac{1}{4^m}.$$

We reason below for the case when E is a finite union of cubes. If E is a finite union of k dimensional smooth manifolds, $m \leq k < n$, then the reasoning below is totally similar, but Lebesgue measure H^n should be replaced by $H^k|E$.

Let H^n denote Lebesgue measure in $\mathbb{R}^n$. The size of $B_0 = B(x_0, R_0)$, $R_0 := \frac{\varepsilon_1}{10}$ centered at x_0 was already very small: at most $\varepsilon_1 \leq \tau_0$. The number τ_0 was chosen in (5.23); it depends on E. So does ε_1. Let us choose ε_1 so much smaller than τ_0 that it also satisfies

$$(5.32) \qquad H^n(B_0 \cap E) \geq a(n)\, R_0^n =: a\, R_0^n.$$

This is possible because of the geometric assumption on E to be a finite union of cubes. Of course ε_1 depends very much on E. But our goal is to guarantee that constants α_0, α_1 can be chosen independent of E. Let us continue. Using (5.31) and Vitali's lemma, we get the covering $\{B(y_j, 3R_j)\}$ of $(B_0 \cap E) \setminus G$ such that $\{B(y_j, r_j^m)\}$ are disjoint and satisfy (5.31). We can shrink r_j to ensure $r_j \leq 5 R_0$ because $\operatorname{supp} \sigma_{00} \subset 2 B_0 = 2 B(x_0, R_0)$. Then

$$H^n((B_0 \cap E) \setminus G) \leq A \sum r_j^n \leq A R_0^{n-m} \sum r_j^m \leq A 4^m R_0^{n-m} \sum \sigma_{00}(B(y_j, r_j)) \leq$$
$$A R_0^{n-m} \sigma_{00}(2 B_0) \leq A 2^m R_0^{n-m} R^m \alpha_1 \leq A R_0^n \alpha_1.$$

We used assumption (5.29). To continue we use (5.32):

$$H^n((B_0 \cap E) \setminus G) \leq \frac{A}{a} \alpha_1 H^n(B_0 \cap E).$$

Constants A, a depend only on n. Thus we can choose $\alpha_1, 0 < \alpha_1 < (80)^{-m}$ and depending only on the dimension in such a way that

$$(5.33) \qquad H^n(G) = H^n(\{y \in B_0 \cap E : M\sigma_0(y) \leq \frac{1}{4^m}\}) > 0.$$

We will prove now that for all sufficiently small positive λ, measure $\sigma_0 + \lambda \cdot H^n|G$ belongs to $\Sigma(E)$:

$$(5.34) \qquad M(\sigma_0 + \lambda \cdot H^n|G) \leq 1 \text{ on } E \text{ if } \lambda \in (0, \lambda_0).$$

In fact, put $\sigma_\lambda := \sigma_0 + \lambda \cdot H^n|G$, fix $z \in E, r > 0$ and consider three cases: 1) $B(z,r) \cap G = \emptyset$, 2) $z \in G$, 3) $z \notin G$, but $B(z,r) \cap G \neq \emptyset$.

In the first case

$$\frac{\sigma_\lambda(B(z,r))}{r^m} = \frac{\sigma_0(B(z,r))}{r^m} \leq 1,$$

as $\sigma_0 \in \Sigma$.

In the second case

$$
\begin{aligned}
\frac{\sigma_\lambda(B(z,r))}{r^m} &= \frac{\sigma_0(B(z,r))}{r^m} + \lambda \frac{H^n(B(z,r) \cap G)}{r^m} \\
&= \frac{\sigma_0(B(z,r))}{r^m} \le \frac{1}{4^m} + \lambda \frac{H^n(B(z,r) \cap B_0)}{r^m} \\
&\le \frac{1}{4^m} + \lambda \cdot \min(r^n, R_0^n)/r^m \le \frac{1}{4^m} + \lambda \cdot R_0^{n-m} < 1 \,,
\end{aligned}
$$

if λ is small enough. (We used the fact that $z \in G$ and the definition of G.)

In the third case, let $y \in B(z,r) \cap G$. Then

$$
\begin{aligned}
\frac{\sigma_\lambda(B(z,r))}{r^m} &\le \frac{\sigma_0(B(z,r))}{r^m} + \lambda \frac{H^n(B(z,r) \cap B_0)}{r^m} \\
&\le 2^m \frac{\sigma_\lambda(B(y,2r))}{2^m r^m} + \lambda \cdot \min(r^n, R_0^n)/r^m \le \frac{1}{2^m} + \lambda \cdot R_0^{n-m} < 1 \,,
\end{aligned}
$$

if λ is small enough. (We used the fact that $y \in G$ and the definition of G.)

Inclusion (5.34) is completely proved. Now we are in a position to use Lemma 5.6. Choose the sequence $\varepsilon_1^{(k)} \le \varepsilon_1$ going to zero. For each k set $R_0^{(k)} = \frac{\varepsilon_1^{(k)}}{10}$, and build a set G_k in $B(x_0, R_0^{(k)})$ as before. Let $H_k := H^n|G_k$. We can apply Lemma 5.6 to $\sigma_0 + \lambda \cdot H_k$. Then we get (see (5.26))

$$
\int \left[|S_{\sigma_0} \mathbf{1}|^2 - 2\Re S_{\sigma_0} S_{\sigma_0} \mathbf{1} \right] \frac{dH_k}{H_k(E)} \ge 1 \,.
$$

Measures $\frac{dH_k}{H_k(E)}$ converge weakly to a point mass measure at x_0. Continuity of kernel of S now gives

(5.35) $$ |S_{\sigma_0} \mathbf{1}|^2(x_0) - 2\Re S_{\sigma_0} S_{\sigma_0} \mathbf{1}(x_0) \ge 1 \,. $$

Let us recall the definition stated before.

Definition. Recall that S has a vector valued kernel,

$$ S_\sigma^* f := S_\sigma^{1,*} f + \cdots + S_\sigma^{d,*} f \,. $$
$$ S_\sigma^* S_{\sigma_0} f := S_\sigma^{1,*}(\overline{S_{\sigma_0}^1 f}) + \cdots + S_\sigma^{d,*}(\overline{S_{\sigma_0}^d f}) \,. $$

Remark. Of course we can forget about complex conjugates if we deal with Lipschitz harmonic capacities.

In these notation, (5.35) implies

(5.36) $$ |S_{\sigma_0}^* \mathbf{1}|^2(x_0) + S_{\sigma_0}^* S_{\sigma_0} \mathbf{1}(x_0) \ge 1/2 \,. $$

This inequality was obtained under the assumption (5.29) that maximal function $M\sigma_0(x_0)$ is smaller than α_1 (and α_1 was chosen to depend only on the dimension to bring us to (5.36)). Put

$$ \alpha_0 := \min(\alpha_1, 1/4) \,. $$

Then Theorem 5.7 is completely proved. $\qquad\square$

Before proceeding we need one more notation.

Notation. Let $\varphi = (\varphi^1, \ldots, \varphi^d)$ be a vector function. Then

$$ S_\sigma^* \varphi := S_\sigma^{1,*}(\varphi^1) + \cdots + S_\sigma^{d,*}(\varphi^d) \,. $$

Denote $\varphi_{\sigma_0} := (\overline{S_{\sigma_0}^1 \mathbf{1}}, \ldots, \overline{S_{\sigma_0}^d \mathbf{1}})$. Then

$$S_\sigma^* \varphi_{\sigma_0} = S_\sigma^* S_{\sigma_0} \mathbf{1}$$

according to the previous definition.

Construction of $Q_1, \ldots, Q_N$. Now we are ready to construct the approximation $\cup_{i=1}^N Q_i$ to E required in (3.13)-(3.15). The idea is that we build the extremal measure σ_0, whose mass is equivalent to $\gamma_{op}(E)$ and whose potential U^{σ_0} "peaks" on E in the sense of lower estimate (5.28). We consider now the "potential neighborhood" of E, namely the set where U^{σ_0} still has a certain estimate from below like (5.28). Then the Whitney decomposition of this neighborhood provides us with the needed cubes Q_i. Unfortunately, to carry out this scheme we need another potential, a bit more complicated than U. Here it is (using the above notation):

$$\mathcal{U}^{\sigma_0}(x) := M\sigma_0(x) + (S_{\sigma_0}^* \mathbf{1})(x) + S_{\sigma_0}^* \varphi_{\sigma_0}(x) + M(|\varphi_{\sigma_0}| \, d\sigma_0)(x).$$

In other words

$$\mathcal{U}^{\sigma_0}(x) = U^{\sigma_0}(x) + M(|\varphi_{\sigma_0}| \, d\sigma_0)(x).$$

Let α_0 be from (5.28). Consider a very small $\beta > 0$ depending only on n and to be chosen later, and set

$$\mathcal{G} := \{y \in \mathbb{R}^n : \mathcal{U}^{\sigma_0}(y) > \beta\alpha_0\}.$$

Let $\{Q^i\}$ denote all Whitney cubes of $\mathcal{G}$. They have disjoint interiors,

$$20\, Q^i \subset \mathcal{G}.$$

(5.37) $$A_4\, Q^i \cap (\mathbb{R}^n \setminus \mathcal{G}) \neq \emptyset.$$

$$\sum \chi_{10\, Q^i} \leq A_5.$$

Choose a finite covering of E from those Q^i that $1.2\, Q^i \cap E \neq \emptyset$. Enumerate them:

$$Q_1, \ldots, Q_n.$$

Put

$$F := 2\, Q_1 \cup \cdots \cup 2\, Q_n.$$

F is a compact subset of $\mathcal{G}$ containing E.

The proof of (3.13). We have to prove only that

(5.38) $$\mathcal{U}^{\sigma_0}(x) \geq \beta\alpha_0 \text{ on } F \Rightarrow \gamma_{op}(F) \leq \frac{C}{\beta\,\alpha_0} \|\sigma_0\|.$$

Choose $\sigma \in \Sigma(F)$ which "almost" gives $\gamma_{op}(F)$ (for example, $\sigma(F) \geq \frac{1}{2}\gamma_{op}(F)$). It satisfies

$$\|S_\sigma : L^2(\sigma) \to L^2(\sigma)\| \leq 1.$$

We will be using repeatedly that then (for arbitrary $i = 1, \ldots d$ and arbitrary measure σ_0; of course we will use it with *our extremal fixed σ_0*)

(5.39) $$\sigma(\{x : S_{\sigma_0}^{i,*}\mathbf{1}(x) \geq t\}) \leq C\,\frac{1}{t}\|\sigma_0\|.$$

Here C depends only on n, m, and Calderón–Zygmund constants of the kernel. We refer the reader to Section 5.3 for this inequality. This is a classical weak type inequality for homogeneous spaces. But 5.39 has been also proved for all CZ operators of order m with respect to any measure of order m. The reader will find the proof in Chapter 5.3.

To prove (5.38), we use that $\mathcal{U}^{\sigma_0}(x) \geq \beta\alpha_0$ on F in the following way. We write first that $\mathcal{U}^{\sigma_0}(x) = M\sigma_0(x) + (S_{\sigma_0}^* \mathbf{1})(x) + S_{\sigma_0}^* \varphi_{\sigma_0}(x) + M(|\varphi_{\sigma_0}| \, d\sigma_0)(x)$, and we can conclude that one of the $2d+2$ terms in this sum is larger than $\frac{\beta\alpha_0}{2d+2}$ on a set $F_0 \subset F$ of measure $\sigma(F_0) \geq \frac{\sigma(F)}{2d+2}$.

Case 1: $M\sigma_0 \geq a := \frac{\beta\alpha_0}{2d+2}$ on F_0 such that $\sigma(F_0) \geq \frac{\sigma(F)}{2d+2}$. For each $x \in F_0$ choose a ball $B(x, r_x)$ such that

$$\sigma_0(B(x, r_x)) \geq a\, r_x^m \,.$$

Use the lemma of Besicovitch to get a finite multiplicity family of $B_j := B(x_j, r_j^m)$ such that it covers F_0. Then (we use that $\sigma \in \Sigma$)

$$\frac{\sigma(F)}{2d+2} \leq \sigma(F_0) \leq \sum \sigma(B_j) \leq \sum r_j^m \leq \frac{1}{a} \sum \sigma_0(B_j) \leq \frac{C\,d}{\beta\,\alpha_0} \|\sigma_0\| \,.$$

This gives (5.38).

Case 2: $S_{\sigma_0}^{i,*}\mathbf{1} \geq a := \frac{\beta\alpha_0}{2d+2}$ on F_0 such that $\sigma(F_0) \geq \frac{\sigma(F)}{2d+2}$. By (5.39) we get

$$\frac{\sigma(F)}{2d+2} \leq \sigma(F_0) = \sigma(\{x : S_{\sigma_0}^{i,*}\mathbf{1}(x) \geq a\}) \leq C\frac{1}{a}\|\sigma_0\| \leq \frac{C\,d}{\beta\,\alpha_0}\|\sigma_0\|$$

and (5.38) follows.

Case 3: $M(|\varphi_{\sigma_0}| \, d\sigma_0) \geq a := \frac{\beta\alpha_0}{2d+2}$ on F_0 such that $\sigma(F_0) \geq \frac{\sigma(F)}{2d+2}$. We repeat verbatim our first case to get

$$\frac{\sigma(F)}{2d+2} \leq \sigma(F_0) \leq \frac{C\,d}{\beta\,\alpha_0}\||\varphi_{\sigma_0}| \, d\sigma_0\| \,.$$

But

$$(5.40) \qquad \||\varphi_{\sigma_0}| \, d\sigma_0\| = \int |S_{\sigma_0}\mathbf{1}| \, d\sigma_0 \leq \|\sigma_0\|^{1/2} (\int |S_{\sigma_0}\mathbf{1}|^2 \, d\sigma_0)^{1/2} \leq \|\sigma_0\| \,,$$

by (5.25). Combining the last two inequalities we get (5.38) in this case too.

Case 4: $S_{\sigma_0}^{i,*}\varphi_{\sigma_0}^i \geq a := \frac{\beta\alpha_0}{2d+2}$ on F_0 such that $\sigma(F_0) \geq \frac{\sigma(F)}{2d+2}$. By (5.39) we get

$$\frac{\sigma(F)}{2d+2} \leq \sigma(F_0) = \sigma(\{x : S_{\sigma_0}^{i,*}\varphi_{\sigma_0}^i(x) \geq a\})$$

$$\leq \frac{C\,d}{\beta\,\alpha_0}\||\varphi_{\sigma_0}^i| \, d\sigma_0\| \leq \frac{C\,d}{\beta\,\alpha_0}\||\varphi_{\sigma_0}| \, d\sigma_0\| \,.$$

Use again that

$$\||\varphi_{\sigma_0}| \, d\sigma_0\| \leq \|\sigma_0\| \,.$$

Combine the last two inequalities. Then (5.38) follows.

Proof of (3.14). We want to choose here β very small but not depending on E to ensure (3.14). Recall that $\varphi_{\sigma_0} := (\overline{S_{\sigma_0}^1 \mathbf{1}}, \ldots, \overline{S_{\sigma_0}^d \mathbf{1}})$. Then

$$S_\sigma^* \varphi_{\sigma_0} = S_\sigma^* S_{\sigma_0}\mathbf{1} \,.$$

Consider $\sigma_j := \sigma_0|4\,Q_j, j+1, \ldots. N$, and new "local" potentials

$$\mathcal{W}^{\sigma_i}(x) := M\sigma_i(x) + (S_{\sigma_i}^* \mathbf{1})(x) + S_{\sigma_i}^* \varphi_{\sigma_0}(x) + M(|\varphi_{\sigma_0}| \, d\sigma_i)(x) \,.$$

LEMMA 5.8. *Let α_0 be from (5.28). There exists $\alpha_3 > 0$ depending only on n, m, d and Calderón–Zygmund constants of the kernel such that*

$$(5.41) \qquad \mathcal{W}^{\sigma_i}(x) \geq \alpha_3\alpha_0 \;\; \forall x \in 3Q_i \cap E \,.$$

Proof. Fix $x \in 3Q_j \cap E$. By (5.37) we fix $z \in A_4 Q_j \cap (\mathbb{R}^n \setminus \mathcal{G})$. In particular,

$$(5.42) \qquad\qquad M\sigma_0(z) \le \beta\alpha_0 \,.$$

$$(5.43) \qquad\qquad S^*_{\sigma_0}\mathbf{1}(z) \le \beta\alpha_0 \,.$$

Then it is easy to see from (5.42), (5.43) that ($R := 2\,A_4\,\mathrm{diam}(Q_j)$)

$$(5.44) \qquad\qquad S^{R,*}_{\sigma_0}\mathbf{1}(x) = \sum_{i=1}^{d} \sup_{r>R} S^{i,r}_{\sigma_0}\mathbf{1}(x) \le \beta\alpha_0 \,.$$

We can also notice that

$$(5.45) \qquad \exists \tilde{A} := A(n,m) \text{ such that } M\sigma_0(x) = \max(\tilde{A}\,\beta\,\alpha_0, M\sigma_j(x)) \,.$$

This is easy:

$$\frac{\sigma_0(B(x,r))}{r^m} \le A\,\frac{\sigma_0(B(z, A'\,A_4\,r))}{r^m} \le A\,(A'\,A_4)^m\,\beta\alpha_0 \,,$$

if $r \ge \frac{\ell(Q_j)}{20}$. Put $\tilde{A} := A\,(A'\,A_4)^m$ in this inequality. Then, calculating $M\sigma_0(x)$, we consider the cases 1) $r \ge \frac{\ell(Q_j)}{20}$, 2) $r < \frac{\ell(Q_j)}{20}$. We have already considered the first case. If the second case happens, then we use the fact that $x \in 3Q_j$, $\sigma_j = \sigma_0|4\,Q_j$, and $r < \frac{\ell(Q_j)}{20}$ to conclude that $\frac{\sigma_0(B(x,r))}{r^m} = \frac{\sigma_j(B(x,r))}{r^m}$. So (5.45) is proved.

But we always have (5.28):

$$M\sigma_0(x) + S^*_{\sigma_0}\mathbf{1}(x) + S^*_{\sigma_0}S_{\sigma_0}\mathbf{1} \ge \alpha_0 \text{ everywhere on } E \,.$$

So one of the $2d+1$ terms of this sum must be at least $\frac{\alpha_0}{2d+1}$. Consider cases.

Case 1: $S^{i,*}_{\sigma_0}\mathbf{1}(x) \ge \frac{\alpha_0}{2d+1}$. Then from (5.44) we conclude ($R := 2\,A_4\,\mathrm{diam}(Q_j)$)

$$S^{i,*}_{\sigma_0|B(x,R)}\mathbf{1}(x) \ge \frac{\alpha_0}{2d+1} - \beta\alpha_0 \ge \frac{\alpha_0}{4d} \,,$$

if β is chosen accordingly. However,

$$S^{i,*}_{\sigma_0|B(x,R)\setminus 4\,Q_j}\mathbf{1}(x) \le C\,\frac{\sigma_0(B(z,2R))}{\ell(Q_j)^m} \,.$$

This is just the rough estimate of the Calderón–Zygmund kernel of S. Taking into account $R \asymp \ell(Q_j)$ and (5.42), we get

$$S^{i,*}_{\sigma_0|B(x,R)\setminus 4\,Q_j}\mathbf{1}(x) \le C\,A\,\beta\alpha_0 \,.$$

Thus

$$S^{i,*}_{\sigma_j}\mathbf{1}(x) = S^{i,*}_{\sigma_0|4\,Q_j}\mathbf{1}(x) \ge \frac{\alpha_0}{4d} - C\,A\,\beta\alpha_0 \ge \frac{\alpha_0}{5d} \,,$$

if β is chosen accordingly. So in our first case (5.41) is proved.

Case 2: $M\sigma_0(x) \ge \frac{\alpha_0}{2d+1}$. Then use (5.45):

$$\frac{\alpha_0}{2d+1} \le M\sigma_0(x) = \max(\tilde{A}\,\beta\,\alpha_0, M\sigma_i(x)) \,.$$

If β is chosen to be smaller than $\frac{1}{2(2d+1)\tilde{A}}$, then this implies

$$M\sigma_i(x) \ge \frac{\alpha_0}{2d+1} \,,$$

and (5.41) is proved too.

Case 3: $S^{i,*}_{\sigma_0}S^i_{\sigma_0}\mathbf{1}(x) = S^{i,*}_{\sigma_0}(\varphi^i_{\sigma_0})(x) \ge \frac{\alpha_0}{2d+1}$.

To consider this last case, notice that we have fixed $x \in 3\,Q_j \cap E$, $z \in A_4\,Q_j \cap (\mathbb{R}^n \setminus \mathcal{G})$ such that

$$(5.46) \qquad M(|\varphi_{\sigma_0}|\,d\sigma_0)(z) \leq \beta\alpha_0$$

and also

$$(5.47) \qquad S^*_{\sigma_0}\varphi_{\sigma_0}(z) \leq \beta\alpha_0\,.$$

Then it is easy to see from (5.46), (5.47) that $(R := 2\,A_4\,\mathrm{diam}(Q_j),\ i = 1, \ldots, d)$

$$(5.48) \qquad S^{i,R,*}_{\sigma_0}(\varphi^i_{\sigma_0})(x) \leq A\,\beta\,\alpha_0\,.$$

Also

$$S^{i,*}_{\sigma_0|B(x,R)\setminus 4\,Q_j}(\varphi^i_{\sigma_0})(x) \leq C\,\frac{\int_{B(z,2R)}|\varphi_{\sigma_0}|\,d\sigma_0}{\ell(Q_j)^m}\,.$$

This is just the rough estimate of the Calderón–Zygmund kernel of S (and the fact that $B(x,R) \subset B(z,2R)$). Taking into account $R \asymp \ell(Q_j)$ and (5.46), we get

$$S^{i,*}_{\sigma_0|B(x,R)\setminus 4\,Q_j}(\varphi^i_{\sigma_0})(x) \leq C\,A\,\beta\alpha_0\,.$$

Thus, using this inequality, (5.48), and the fact that

$$S^{i,*}_{\sigma_0}(\varphi^i_{\sigma_0})(x) \leq S^{i,*}_{\sigma_0|4\,Q_j}(\varphi^i_{\sigma_0})(x) + S^{i,*}_{\sigma_0|B(x,R)\setminus 4\,Q_j}(\varphi^i_{\sigma_0})(x) + S^{i,R,*}_{\sigma_0}(\varphi^i_{\sigma_0})(x)\,,$$

we get

$$S^{i,*}_{\sigma_j}(\varphi^i_{\sigma_0})(x) = S^{i,*}_{\sigma_0|4\,Q_j}(\varphi^i_{\sigma_0})(x) \geq \frac{\alpha_0}{2d+1} - C\,A\,\beta\alpha_0 \geq \frac{\alpha_0}{4d}\,,$$

if β is chosen accordingly. So in our third case (5.41) is proved as well. Lemma 5.8 is completely proved. $\qquad\qquad\square$

From Lemma 5.8 it is easy now to prove (3.14). In fact, recall (5.38). In our case Lemma 5.8 says that

$$\mathcal{W}^{\sigma_j}(x) := M\sigma_j(x) + (S^*_{\sigma_j}\mathbf{1})(x) + S^*_{\sigma_j}\varphi_{\sigma_0}(x) + M(|\varphi_{\sigma_0}|\,d\sigma_j)(x) \geq \alpha_3\,\alpha_0 \ \text{ on } 3\,Q_j \cap E\,.$$

We can repeat the proof of (5.38) verbatim to obtain

$$(5.49) \quad \mathcal{W}^{\sigma_j} \geq \alpha_3\,\alpha_0 \text{ on } 3\,Q_j \cap E \Rightarrow \gamma_{op}(3\,Q_j \cap E) \leq \frac{C}{\alpha_3\,\alpha_0}(\|\sigma_j\| + \|\varphi_{\sigma_0}\,d\sigma_j\|)\,.$$

But

$$(5.50) \qquad \sum \|\sigma_j\| + \|\varphi_{\sigma_0}\,d\sigma_j\| \leq A\,(\|\sigma_0\| + \int_{4\,Q_j}|\varphi_{\sigma_0}|\,d\sigma_0) \leq A\,\|\sigma_0\|\,.$$

This is by the finite multiplicity of Q_j and by (5.40).

Combining (5.50), (5.49), we get

$$(5.51) \qquad \sum_j \gamma_{op}(3\,Q_j \cap E) \leq A\,\gamma_{op}(E)\,.$$

Now (3.14) follows from

$$(5.52) \qquad \sum_j \gamma_+(3\,Q_j \cap E) \leq A\,\gamma_+(E)\,,$$

And this follows from (5.51) and the following equivalence, which will be proved in Section 5.4 below:

$$(5.53) \qquad A^{-1}\gamma_+(E) \leq \gamma^*_+(E) \leq \gamma_+(E)\,.$$

We already know that γ_{op} and γ^*_+ are equivalent; see (5.10), (5.11), (5.15) combined.

Proof of (3.15). We work under assumption (5.2). Notice that it is obvious that (we use (5.40) here)

$$\mathcal{U}^{\sigma_0}(x) \leq \frac{\|\sigma_0\|}{\operatorname{dist}(x, E)^m}\,.$$

Now use $m = n - 1$ and (5.2). From this, for $x \in \partial\mathcal{G}$ (recall that $\|\sigma_0\| \leq A_2 \gamma_+^*(E)$),

$$\operatorname{dist}(x, E)^{n-1} \leq A_1\, A_2\, \frac{1}{\beta\,\alpha_0}\gamma_+^*(E) \leq A_1\, A_2\, \frac{1}{\beta\,\alpha_0}\, A_3\gamma_+(E) \leq$$

$$A_1\, A_2\, \frac{1}{\beta\,\alpha_0}\, A_3\, \frac{1}{A_+\, A}\, \operatorname{diam}(E)^{n-1}\,.$$

Choosing A_+ in (5.2) to kill all of these constants we clearly have (3.15) because the boundary of $\mathcal{G}$ becomes, say $1/1000\,\operatorname{diam}(E)$ close to E. We used in the chain above the following inequality:

(5.54) $$\gamma_+^*(E) \leq A_3\, \gamma_+(E)\,.$$

It is just one part of (5.53), which will be proved in Section 5.4 below.

5.3. L^p theory of nonhomogeneous CZ operators. Measure of order m

In this section T is an operator with vector valued Calderón–Zygmund kernel of order m; see (8.7). Measure μ is always assumed to be of order m, too; see (10.9).

We may use the notation $T \in CZ(K)$ to say that T is a Calderón-Zygmund operator with kernel K. We need also the notions of "cut-off" of T.

$$T^\varepsilon f(x) := \int_{y:|y-x|\geq\varepsilon} K(x,y)f(y)\, d\mu(y),$$

$$T^{(\varepsilon)} f(x) := \int K(x,y)\psi\big(\frac{|x-y|}{\varepsilon}\big)f(y)\, d\mu(y),$$

where ψ is a C^∞ function that vanishes on $B(0, 1/2)$ and equals 1 on $\mathbb{R}^n \setminus B(0, 1)$. Notice that $T^{(\varepsilon)}$ are operators with Calderón-Zygmund kernels, while T^ε are not.

The "right" maximal operator is now

$$M_\mu f(x) := \sup_r \frac{1}{\mu(B(x, 3r))} \int_{B(x,r)} |f(y)|\, d\mu(y)\,.$$

Obviously,

(5.55) $$|(T^\varepsilon - T^{(\varepsilon)})(f)(x)| \leq AM_\mu f(x)\,.$$

We also introduce the *singular maximal function*:

$$T^* f(x) := \sup_\varepsilon |T^\varepsilon f(x)|\,.$$

The next theorem is important in what follows. See [**44**] for the proof.

THEOREM 5.9. *If $T \in CZ(K)$, then any of the following equivalent assertions hold:*

1) $\{T^\varepsilon\}_{\varepsilon>0}$ are uniformly bounded in $L^2(\mu)$,
2) $\{T^{(\varepsilon)}\}_{\varepsilon>0}$ are uniformly bounded in $L^2(\mu)$,
3) T^* is bounded in $L^2(\mu)$.

The proof can be found in [**43**], [**46**].
Let $M_0(\mathbb{R}^n)$ stand for all complex measures with compact support in $\mathbb{R}^n$.

THEOREM 5.10. *If $T \in CZ(K)$, then*

1) T is a bounded operator from $L^1(\mu)$ to $L^{1,\infty}(\mu)$.

2) If in addition the kernel K is continuous, then T is a bounded operator from $M_0(\mathbb{R}^n)$ to $L^{1,\infty}(\mu)$ and its norm depends only on the Calderón-Zygmund parameters z and the norm of T in $L^2(\mu)$.

The proof can be found in [**44**].

The next lemma is well known and widely used; see [**5**].

LEMMA 5.11. *Let $\mathcal{X}$ be a compact Hausdorff space and $T = (T_1, \ldots, T_d)$ be a bounded linear operator from $M(\mathcal{X})$ to $C(\mathcal{X}) \times C(\mathcal{X}) \times \ldots C(\mathcal{X})$ ($d < \infty$ times), where $M(\mathcal{X})$ is the space of complex measures on $\mathcal{X}$. Also assume that the adjoint operators $T_i^{'}$ acts from $M(\mathcal{X})$ to $C(\mathcal{X})$, and that for a finite positive measure μ, the following holds:*

$$(5.56) \qquad \mu\{x \in \mathcal{X} : |T_i^{'}\nu(x)| > t\} \leq \frac{A\|\nu\|}{t} \quad \forall \nu \in M(\mathcal{X}).$$

Then for any Borel set $E \subset \mathcal{X}$, $0 < \mu(E)$, there exists a function h on E, $0 \leq h \leq 1$, such that

$$(5.57) \qquad \int_E h \, d\mu \geq \frac{\mu(E)}{2},$$

$$(5.58) \qquad \|T(h \, d\mu)\|_\infty < 4\,A.$$

The proof can be found in [**68**], [**5**].

THEOREM 5.12. *Let $T \in CZ(K)$, with $\|T\|$ denoting its norm as an operator in $L^2(\mu)$. Let $\delta \in (0,1)$. Then there exist constants C_1, C_2 depending on δ and $\|T\|$ only such that*

$$(5.59) \qquad T^*f(x) \leq C_1(M_\mu|Tf|^\delta)^{1/\delta}(x) + C_2(M_\mu f)(x), \quad \forall f \in L^1(\mu).$$

The same is true for the operator with vector valued kernel. Also then T^* is a bounded operator from $M_0(\mathbb{R}^n)$ to $L^{1,\infty}(\mu)$ and its norm depends only on the Calderón-Zygmund parameters z and the norm of T in $L^2(\mu)$.

The last statement was repeatedly used in Section 5.2. The reader can find the proof in [**44**]. But notice that a more general theorem is proved below in Chapter 11. We say " more general" because in Chapter 11, even the assumption that μ is of order m is omitted.

This is called Cotlar's inequality. Notice that $|Tf|^\delta$ is summable by Theorem 5.10. This makes the right side almost everywhere finite. The proof can be found in [**44**].

Theorem 5.13 can be found in [**45**]. But we decided to give its proof because it was used in an important fashion when we compared different capacities with Calderón–Zygmund kernels at the beginning of this section.

THEOREM 5.13. *Let μ be supported by a compact set E, and let an operator T with Calderón-Zygmund kernel be bounded in $L^2(\mu)$. Then there exist $\delta > 0$ and $B < \infty$ which depend only on $\|T\|$, and a function b, such that*

$$(5.60) \qquad \|b\|_\infty \leq 1,$$

$$(5.61) \qquad \Re \int_E b \, d\mu \geq \delta\,\mu(E),$$

$$(5.62) \qquad \|T(b \, d\mu)\|_\infty \leq B.$$

Proof. We deduce from Theorems 5.9, 5.10 that

$$(5.63) \qquad (T^{(\varepsilon)})^* : M(E) \to L^{1,\infty}(\mu)$$

uniformly in ε.

Obviously $T^{(\varepsilon)}$ satisfies all the assumptions of Lemma 5.11. Applying this lemma we obtain the family of functions $\{h^\varepsilon\}$ having the following properties:

$$0 \leq h^\varepsilon \leq 1,$$

$$\int_E h^\varepsilon \, d\mu \geq \delta \mu(E),$$

$$\|T^{(\varepsilon)}(h^\varepsilon \, d\mu)\|_\infty \leq B.$$

Let us consider a decreasing sequence $\varepsilon_n \to 0$ such that $h^{\varepsilon_n} \to h$ weakly in $L^2(\mu)$ and $T^{(\varepsilon_n)} \to T_0$ in the weak operator topology (we can do that because the $T^{(\varepsilon_n)}$ are uniformly bounded operators from a separable to a reflexive space). It is clear that $T_0 \in CZ(K)$. Let $h_n := h^{\varepsilon_n}$, $T_n := T^{\varepsilon_n}$.

LEMMA 5.14. *The functions $T_n h_m$, $m \geq n$, are uniformly bounded.*

Proof. We know that $\|T^{(\varepsilon)}(h^\varepsilon \, d\mu)\|_\infty \leq B$. Using (5.55) and the uniform boundedness of h_m we conclude that $\|T_n h_m\| \leq B_1$. Similarly to (5.55)

$$|T^\delta f(x) - T^{2\delta} f(x)| \leq A \, M_\mu f(x).$$

Thus, if $\varepsilon_n \in [\varepsilon_m, 2\varepsilon_m]$, then $\|T_n h_m\|_\infty \leq B_2$. If $\varepsilon_n > 2\varepsilon_m$, then $(T^{(\varepsilon_m)})^{\varepsilon_n} = T^{\varepsilon_n}$. Thus, in this case

$$(5.64) \qquad \|T_n h_m\|_\infty = \|(T^{(\varepsilon_m)})^{\varepsilon_n} h^{\varepsilon_m}\|_\infty \leq \|(T^{(\varepsilon_m)})^* h^{\varepsilon_m}\|_\infty.$$

By Cotlar's inequality (Theorem 5.58) we get

$$(5.65) \qquad ((T^{(\varepsilon_m)})^* h^{\varepsilon_m}(x) \leq C_1(M_\mu|(T^{(\varepsilon_m)})^{\varepsilon_n} h^{\varepsilon_m}|^\delta)^{1/\delta}(x) + C_2(M_\mu h^{\varepsilon_m})(x).$$

But both h^{ε_m} and $(T^{(\varepsilon_m)})^{\varepsilon_n} h^{\varepsilon_m}$ are uniformly bounded. Thus (5.64) and (5.65) give us finally the uniform boundedness of $T_n h_m$, for all $m \geq n$. $\qquad \square$

To continue the proof of Theorem 5.13 let us choose $\{a_k^n\}_{k=0}^{m_n}$, $a_k^n \geq 0$, $\Sigma_k a_k^n = 1$, in such a way that $g_n := \Sigma_{j=0}^{m_n} a_j^n h_{n+j}$ converges in $L^2(\mu)$ norm to an h. Consider $T_n g_n$. Then

$$\|T_n g_n\|_\infty \leq (\Sigma_{j=0}^{m_n} a_j^n) \sup_{m \geq n} \|T_n h_m\|_\infty \leq A < \infty.$$

In particular, let n_j be a sequence such that

$$(5.66) \qquad T_{n_j} g_{n_j} \to f \quad \text{weakly in } L^2(\mu).$$

Clearly f is a bounded function and

$$(5.67) \qquad \|f\|_\infty \leq A := \limsup_{n \to \infty} \|T_n g_n\|_\infty < \infty.$$

Let us prove that $T h$ is bounded. This will finish the proof of the theorem because

$$(5.68) \qquad \int_E h \, d\mu = \lim_n \int_E g_n \, d\mu = \lim_n (\Sigma_{j=0}^{m_n} a_j^n \int_E h_{n+j} \, d\mu) \geq \delta \mu(E).$$

To show that $T h$ is bounded, we will show first that $T_0 h = f$. In fact, $T_0 h = T_0 h - T_{n_j} g_{n_j} + T_{n_j} g_{n_j} = T_{n_j} (h - g_{n_j}) + (T_0 - T_{n_j}) h + T_{n_j} g_{n_j}$.

The first term tends to zero in $L^2(\mu)$ because the T_{n_j} are uniformly bounded and $g_{n_j} \to h$ in $L^2(\mu)$. The second term tends weakly to zero because we have chosen T_n converging in the weak operator topology to T_0. By (5.66) the third term tends weakly to f. So $T_0 h = f$. But T and T_0 are both from $CZ(K)$; and they are both bounded in $L^2(\mu)$. Thus $T h - T_0 h = m h$, where $m \in L^\infty(\mu)$. Thus $T h = m h + f$, and so it is a bounded function. We also guaranteed (5.68). The construction implies $\|h\|_\infty \le 1$ obviously.

Theorem 5.13 is completely proved. $\qquad\square$

5.4. Riesz and Cauchy kernels: $\gamma_+ \asymp \gamma_{op}$

We need to compare two capacities: γ_+ introduced in Chapter 2 and γ_+^* introduced in Section 5.1. Let

$$k(x,y) : \mathbb{R}^n \times \mathbb{R}^n \to \mathbb{C}^d$$

be a vector valued Calderón–Zygmund kernel of order $m < n$. Let $E \subset \mathbb{R}^n$ be a compact set, and let μ here always denote a positive measure of order m on E. We write the last assumption as

$$\mu \in \Sigma(E).$$

We assign to k and μ the following potentials:

$$T^*\mu(x) = T_\mu^*\mathbf{1}(x) := \sup_{r>0} \left| \int_{y:|y-x|>r} k(x,y)\, d\mu(y) \right| \; x \in E.$$

$$\mathcal{T}^\mu(x) := \int k(x,y)\, d\mu(y) \; x \in \mathbb{R}^n.$$

Remark. $\mathcal{T}^\mu(x)$ is defined a.e. in $\mathbb{R}^n$ with respect to Lebesgue measure H^n of $\mathbb{R}^n$. This is because m (the order of k) is assumed to be strictly less than n.

Recall that $\Phi * \cdot$ denotes the Newton potential, here

$$\Phi(x,y) = c_n \frac{1}{|x-y|^{n-2}}, n > 2; \; = c \log \frac{1}{|x-y|}, n = 2.$$

Recall that we have some special kernels: 1) the Riesz kernel on $\mathbb{R}^n$ has order $n-1$; it is

$$\Psi(x,y) = \Psi_n(x,y) := \left(\frac{x_1 - y_1}{|x-y|^n}, \dots, \frac{x_n - y_n}{|x-y|^n} \right) \; x,y \in \mathbb{R}^n,$$

and 2) the Cauchy kernel on $\mathbb{R}^2$:

$$K(z,\zeta) := \frac{1}{z-\zeta} = \frac{x_1 - y_1}{|x-y|^2} - i\frac{x_2 - y_2}{|x-y|^2}, \quad z = x_1 + ix_2, \; \zeta = y_1 + iy_2 \in \mathbb{C}.$$

Notice that Ψ_2 and K are very close but different.

Here is an obvious

LEMMA 5.15. *Let γ_+ be a capacity introduced in Chapter 2. Let $k = \Psi$, that is the Riesz kernel. Then*

$$(5.69) \qquad \gamma_+(E) \asymp \sup\{\mu(E) : \mu \in \Sigma(E), |\mathcal{T}^\mu(x)| \le 1 \text{ for } H^n \text{ a.e. } x \in \mathbb{R}^n\}.$$

Proof. We just need to see that if μ participates in $\gamma_+(E)$, then μ/A is of order $n-1$. But this follows from Lemma 4.1. In the opposite direction we have to notice that if $\mu \in \Sigma(E), |\mathcal{T}^\mu(x)| \le 1$ for H^n a.e. $x \in \mathbb{R}^n$ with Riesz kernel as a kernel of $\mathcal{T}$, then the Newton potential $\Phi * \mu$ is a Lipschitz function in $\mathbb{R}^n$. But the generalized derivative of the Newton potential $\Phi * \mu$ is exactly $\mathcal{T}^\mu(x)$ considered as a function in $L^1_{loc}(\mathbb{R}^n)$. Then the relationship $|\mathcal{T}^\mu(x)| \leq 1$ for H^n a.e. $x \in \mathbb{R}^n$ says that $\Phi * \mu$ is Lipschitz in $\mathbb{R}^n$.   □

Remark. 1) If $|\mathcal{T}^\mu(x)| \leq 1$ for H^n a.e. $x \in \mathbb{R}^n$, then there exists A depending only on n such that $\mu/A \in \Sigma$. This is clear from Lemma 4.1. It means that we can skip the requirement $\mu \in \Sigma(E)$ in the right part of (5.69).

2) Consider the Cauchy kernel K. Introduce

$$\gamma_+(E) := \sup\{\mu(E) : \mu \in \Sigma(E), |\mathcal{T}^\mu(x)| \leq 1 \text{ for } H^2 \text{ a.e. } x \in \mathbb{C}\}$$

and

$$\Gamma_+(E) := \sup\{\mu(E) : \mu \in \Sigma(E), |\mathcal{T}^\mu(z)| \leq 1 \text{ for every } z \in \mathbb{C} \setminus E\}.$$

Then, of course,

$$\gamma_+(E) \leq \Gamma_+(E), \quad \gamma_+(E) = \Gamma_+(E) \text{ if } H^2(E) = 0.$$

Definition. Analytic capacity of Vitushkin is defined for compact $E \subset \mathbb{C}$ with the help of Cauchy kernel K (here S denote the distributions with compact support on E):

(5.70) $$\Gamma(E) := \sup\{|\langle S, \mathbf{1} \rangle : |K * S(z)| \leq 1 \ \forall z \in \mathbb{C} \setminus E\}.$$

We are in process of proving (and the proof will take 80 pages more) that

(5.71) $$\gamma_+(E) \asymp \Gamma(E).$$

As, by definition, $\Gamma_+(E) \leq \Gamma(E)$, we will get that

$$\gamma_+(E) \asymp \Gamma_+(E),$$

for all E, including those for which $H^2(E) > 0$.

Problem. Find a simple proof of

$$\gamma_+(E) \asymp \Gamma_+(E),$$

for E of positive Lebesgue measure.

Now our task is much simpler. We are going to prove that for a general Calderón–Zygmund kernel of order m introduced at the beginning of this section two capacities are equivalent. Let us use the notation: $T^*\mu(x) := T^*_\mu \mathbf{1}(x)$ for maximal singular operator. Here $T^*_\mu \mathbf{1}(x) := \sup_{r>0} |\int_{y:|y-x|>r} k(x,y) \, d\mu(y)|$. Then

$$\gamma^*_+(E) = \sup\{\mu(E) : \mu \in \Sigma, T^*\mu(x) \leq 1 \ \forall x \in E\},$$

$$\gamma_+(E) = \sup\{\mu(E) : \mu \in \Sigma(E), |\mathcal{T}^\mu(x)| \leq 1 \text{ for } H^n \text{ a.e. } x \in \mathbb{R}^n\}.$$

We abuse the notation slightly because the term $\gamma_+(E)$ was defined differently in Chapter 2. But this is not a problem because by Lemma 5.15 "old" $\gamma_+(E)$ and "new" $\gamma_+(E)$ are equivalent.

THEOREM 5.16. *Let $m < n$, then $\gamma^*_+(E) \asymp \gamma_+(E)$.*

Proof. 1) $\gamma^*_+(E) \leq A \gamma_+(E)$. If μ participates in $\gamma^*_+(E)$, then of course $|\mathcal{T}^\mu(x)| \leq 1$ for H^n a.e. $x \in \operatorname{supp} \mu$. In fact, for a.e. point x with respect to H^n the function $y \to |k(x,y)|$ belongs to $L^1(\mu)$. In particular, the integral involved in $\mathcal{T}^\mu(x)$ converges absolutely. If, in addition, $x \in \operatorname{supp} \mu$, then $T^*_\mu \mathbf{1}(x) \leq 1$. And we conclude $|\mathcal{T}^\mu(x)| \leq 1$ by the Lebesgue Dominant Convergence Theorem. So on $\operatorname{supp} \mu$ we are fine.

Now let μ participates in $\gamma_+^*(E)$, and let $x \in \mathbb{R}^n \setminus E$. Let $z \in \operatorname{supp} \mu$ be a closest point to x. Let $r = |z - x|$. Then $T_\mu^{2r}\mathbf{1}(z) \leq 1$ because $T_\mu^*\mathbf{1}(x) \leq 1$, and

$$|T_\mu^{2r}\mathbf{1}(z) - T^\mu(x)| \leq C \int_{y:|y-z|\geq r} \frac{r^\tau}{|z-y|^{m+\tau}} \, d\mu(y) \leq C \,.$$

This is by the standard Calderón–Zygmund estimate of the kernel (see (8.7)) and by the fact that $\mu \in \Sigma(E)$. Therefore, for μ participating in $\gamma_+^*(E)$ we have $|T^\mu(x)| \leq C + 1$, and our first inequality is proved.

2) $\gamma_+(E) \leq A \gamma_+^*(E)$. Let ψ be a C^∞ function with support in $B(0,1)$, $\int \psi \, dx = 1$. Let ψ_τ denote $\frac{1}{\tau^n}\psi(\frac{\cdot}{\tau})$. If μ participates in $\gamma_+(E)$, then obviously

$$|(\psi_\tau * T^\mu)(x)| \leq 1 \;\; \forall x \in \mathbb{R}^n \,.$$

Let $x \in E$ and $r > 0$. First of all, if $|x - x'| \leq \frac{r}{2}$, then

$$\left| T^r\mu(x) - \int_{y:|y-x|\geq r} k(x',y) \, d\mu(y) \right|$$

$$= \left| \int_{y:|y-x|\geq r} k(x,y) \, d\mu(y) - \int_{y:|y-x|\geq r} k(x',y) \, d\mu(y) \right|$$

$$\leq C \int_{y:|y-z|\geq r} \frac{r^\tau}{|z-y|^{m+\tau}} \, d\mu(y) \leq C \,.$$

Denote $t(x') := \int_{y:|y-x|\geq r} k(x',y) \, d\mu(y)$, $R(x') := \int_{y:|y-x|<r} k(x',y) \, d\mu(y)$. Then we have just proved that if $\tau \leq r/2$ we have

$$|T^r\mu(x) - (\psi_\tau * t)(x)| \leq C \,.$$

We are left to notice that

$$(\psi_{r/2} * t)(x) = (\psi_{r/2} * T^\mu)(x) - (\psi_{r/2} * R)(x) =: I + II$$

and both terms are bounded. In fact, we already noticed that $|I| \leq 1$. To estimate $|II|$ we write (we choose $\tau = r/2$)

$$|II| \leq \int |\psi_{r/2}(x - x')| \, dx' \int_{y:|y-x|<r} |k(x',y)| \, d\mu(y)$$

$$\leq A \frac{1}{r^n} \int_{y:|y-x|<r} d\mu(y) \int_{x':|x'-y|\leq \frac{3r}{2}} \frac{dx'}{|x'-y|^m}$$

$$\leq A \frac{r^{n-m}}{r^n} \int_{y:|y-x|<r} d\mu(y) \leq A \frac{1}{r^m}\mu(B(x,r)) \,.$$

The last quantity is bounded as $\mu \in \Sigma$. Finally, with any $r > 0$ we got

$$|T^r\mu(x)| \leq C \,.$$

So $T^*\mu(x) \leq C$, and our second inequality is proved too. The theorem is completely proved. $\qquad\square$

5.5. Cauchy kernel and analytic capacity

We are in $\mathbb{C} = \mathbb{R}^2$ here. Kernels $k = K$ (Cauchy) and $k = \Psi_2$ (Riesz) are very similar but different. Also the definition of capacities (γ, γ_+ for Riesz potentials, Γ, Γ_+ for Cauchy potential) are quite different. To define analytic capacity the restriction on the Cauchy potential is imposed only in $\mathbb{R}^2 \setminus E$. For Riesz potential the restriction on the potential, which defines the capacities (now Lipschitz harmonic capacity, for example), is imposed a.e. in $\mathbb{R}^2$ with respect to Lebesgue measure. This last distinction can be forgotten by the virtue of the following reasoning. Our goal is to prove the equivalence of $\Gamma(E), \Gamma_+(E)$. By considering approximation to E by piecewise smooth curve we can always think that

(5.72) $\qquad\qquad\qquad E$ is a finite union of smooth curves.

Then there is no difference between the a.e. requirement and the outside of E requirement. Let us call $\mathcal{F}$ a family of compact sets such that if equivalence of two capacities is proved for every compact set $E \in \mathcal{F}$ (with constants independent from E), then it is proved for every compact set. For Γ, Γ_+ one of such families is described in (5.72). For Lipschitz harmonic capacities $\gamma(E), \gamma_+(E)$ the family of sets which are a finite union of cubes can serve as such a family.

Let us try to list what exactly we used from the potential $V = k * \cdot$ (it was always Riesz potential before, $k = \Psi_n$, but now it can be any Calderón–Zygmund kernel k of order m including Cauchy kernel $k = K$), family $\mathcal{F}$ of compact sets E, and capacity $C(E)$ (it was always Lipschitz harmonic capacity $\gamma(E)$ before, but now can be $\Gamma(E)$ too).

Axiom I. $C(E) \le A \,(\mathrm{diam}(E))^m$.

Axiom II. If $C(E) > 0$ and $E \in \mathcal{F}$, then in any neighborhood E_τ of E, one can find a non-zero complex measure ν on E_τ such that

$$|V^\nu| \le 1 \text{ for a.e. } x \in \mathbb{R}^n \text{ with respect to Lebesgue measure on } \mathbb{R}^n.$$

Axiom III. If

$$|V^\nu| \le 1 \text{ for a.e. } x \in \mathbb{R}^n \text{ with respect to Lebesgue measure on } \mathbb{R}^n,$$

and if $\varphi \in C_0^\infty(B(x, R))$, $|\varphi| \le 1, |\nabla^2 \varphi| \le \frac{A}{r^2}$, then

$$|V^{\varphi \nu}| \le 1 \text{ for a.e. } x \in \mathbb{R}^n \text{ with respect to Lebesgue measure on } \mathbb{R}^n.$$

Axiom IV. If the distribution with compact support on $E \in \mathcal{F}$ is such that $\langle S, \mathbf{1} \rangle = 0$, then

$$|V^S(z)| \le \frac{A \,\mathrm{diam}(E)\, \gamma(E)}{\mathrm{dist}(z, E)^{m+1}} \ \forall z \in \mathbb{R}^n \text{ such that } \mathrm{dist}(z, E) \ge \mathrm{diam}(E).$$

All axioms hold for $n = 2, m = 1$, E from (5.72), potential with kernel $k = K$ (=Cauchy kernel $\frac{1}{z-\zeta}$). Let us notice that, maybe, only Axioms II and IV should be addressed here for Cauchy potentials. As to Axiom IV the method of Chapter 3 can be not applicable here. But still, it is well-known that for Cauchy potentials one has the estimate of Axiom IV. See [**21**], for example. The proof is very easy, but it uses complex analysis. Complex analysis is not available for the Riesz potential analog of this inequality. This is why we did the estimate in Chapter 3 for Riesz potentials. As to Axiom II it is readily fulfilled for all E of special class: E is a finite union of smooth curves on the plane. But it is very easy to understand that only for such E's we need to prove the equivalence of different capacities (with constants of equivalence independent of E of course).

So all below and above is applicable. And along with $\gamma(E) \asymp \gamma_+(E)$ for Lipschitz harmonic capacities we are proving the following:

$$\Gamma(E) \asymp \gamma_+(E).$$

This proves simultaneously Tolsa's theorem (it was known as Melnikov's conjecture):

$$\Gamma(E) \asymp \Gamma_+(E),$$

and semiadditivity of Γ (Vitushkin's conjecture). The last claim holds, because we know that $\gamma_+ \asymp \gamma_+^* \asymp \gamma_{op}$. And γ_{op} is obviously semiadditive.

Remark. Of course, there is nothing new in this. Tolsa's theorem already did that. We are just indicating that it can be applied to more general (Lipschitz harmonic, for example) capacities. This is also not so strange as nonhomogeneous Tb theorem on which everything is based (see [46],[47]); it is very general and not kernel specific.

CHAPTER 6

The Tree of the Proof

Fix a compact set $E \subset \mathbb{R}^n$. We use the family $\{Q_i\}_{i=1}^N$ with properties (3.13)-(3.15), the existence of which is justified in the previous section. Let $E_i := 2Q_i \cap E$. Now two cases may occur.

First case:

$$(6.1) \qquad \gamma(E) < C_1^{-1} \sum_{i=1}^N \gamma(E_i) \,.$$

Recall that C_1 is a special name given to the constant from (3.14).

Second case:

$$(6.2) \qquad \gamma(E) \geq C_1^{-1} \sum_{i=1}^N \gamma(E_i) \,.$$

First step is fulfilled. Suppose the first case occurs. The procedure of splitting performed for E can be repeated for its pieces E_i : for every E_k, $k = 1, \ldots, N$. We fix k, think that E_k is E, and repeat the approximation procedure by family $\{Q_{ki}\}$ getting $E_{ki} = 2Q_{ki} \cap E_k$. Let us emphasize that when the first case occurs for $(E, \{E_i\}_{i=1}^N)$, we do this for every $k = 1, \ldots, N$. Now let k be fixed. Consider E_k and $\{E_{ki}\}$. Again two cases may occur. If the first case occurs, we do the splitting of E_{ki} for all i's. If the second case occurs for $(E_k, \{E_{ki}\})$; we stop here, we do not split any of E_{ki}. We call Q_{ki}'s stopping cubes; and we call multi-indices $J = ki$ stopping indices.

Second step is fulfilled. Suppose the second case occurs. We do not split E_i anymore and we will never split them in the future. We call Q_i's stopping cubes. Indices $J = i, i = 1, \ldots, N$ are called stopping indices.

Notice that the nature of the case (first or second) depends on $(E_J, \{E_{Ji}\})$. If the first step occurs for $(E_J, \{E_{Ji}\})$, one then splits E_{Ji} for every i. If the second step occurs for $(E_J, \{E_{Ji}\})$, one does not split any of E_{Ji}. One just calls Q_{Ji} stopping cubes and multi-indices Ji are called stopping indices.

As a result, after finitely many steps we get the collection of E_J, where J is a multi-index; its length varies from 1 to some finite number. All multi-indices except the longest ones are stopping indices. Some longest ones can be stopping indices too.

Let us first assume that only the first case occurs all the time. We have $E = \cup E_J$, where all indices J are of the same length (say ℓ) and by (6.1),

$$(6.3) \qquad \gamma(E) < C_1^{-\ell} \sum_J \gamma(E_J) \,.$$

Suppose now that if ℓ is very large, then there exists a constant $\mathcal{A}$ depending only on dimension such that

(6.4) $$\gamma(E_J) \leq \mathcal{A}\,\gamma_+(E_J)\,.$$

Combine (6.4) with (6.3) to get

(6.5) $$\gamma(E) < \mathcal{A}\,C_1^{-\ell} \sum_J \gamma_+(E_J)\,.$$

Let now (3.14) come into play. Repeatedly using it (in ℓ generations of splitting of E), we get

(6.6) $$\sum_J \gamma_+(E_J) \leq C_1^{\ell}\gamma_+(E)\,.$$

The combination of two previous inequalities gives

(6.7) $$\gamma(E) \leq \mathcal{A}\,\gamma_+(E)\,.$$

This is what we wanted to prove. We postpone a bit the discussion of why (6.4) is valid. Instead, let us first assume its validity and let us get (6.7) in the situation when not only the first case occurs. In this general situation $E = \cup_J E_J$, where J's are of two sorts: either the longest length ℓ or stopping indices (or both). For any stopping index J let $\hat{J}$ mean its "father"; in other words $J = \hat{J}i$. (Notice that $\hat{J}$ can be empty if the second case happens already for $(E, \{E_i\}_{i=1}^N)$.) We cannot write (6.3) because the second case occurs in passing from $\hat{J}$ to its "sons" J, and this is exactly the place where the opposite inequality (6.2) takes place. But if we stop every time at "fathers" of stopping indices we can write the analog of (6.3):

(6.8)
$$\gamma(E) < \sum_{m=0}^{\ell-1} C_1^{-m} \sum_{\hat{J}:\text{length of } \hat{J}=m} \gamma(E_{\hat{J}}) + C_1^{-\ell} \sum_{J:\text{length of } J=\ell,\, J \text{ is not stopping}} \gamma(E_J)\,.$$

In the second sum one can use (6.4) again if ℓ is very large (we will validate (6.4) for large ℓ very soon). But in the first sum this validation will not work because index m generally is not large at all. Here one needs to use something else. In the first sum we use the fact that for every "father" $\hat{J}$ of stopping indices $\{Ji\}$, one has

(6.9) $$\gamma(E_{\hat{J}}) \leq \mathbf{A}\,\gamma_+(E_{\hat{J}})\,,$$

where $\mathbf{A}$ depends at worst on the dimension.

The last inequality is the most difficult one. It follows from the theorem below. The proof of this theorem will occupy practically the rest of our consideration.

THEOREM 6.1. *Let E be a compact in $\mathbb{R}^n$, and let $\{Q_i\}_{i=1}^N$ be cubes with properties (3.13)-(3.14). Let (6.2) occur. Then there are constants α, β, λ (depending at worst on the dimension) such that*

(6.10) $$\gamma(E) \leq \lambda C_0^{\alpha}\,C_1^{\beta}\,\gamma_+(E)\,.$$

Taking Theorem 6.1 for granted, we can put $\mathbf{A} = \lambda C_0^{\alpha}\,C_1^{\beta}$ and get (6.9) by applying Theorem 6.1 to $E_{\hat{J}}, length(\hat{J}) = m \leq \ell - 1$. Let $A = \max(\mathbf{A}, \mathcal{A})$. Now

(6.9) (used in the first sum of (6.8)) and (6.4) (used in the second sum of (6.8)) combined give

(6.11)
$$\gamma(E) < A \sum_{m=0}^{\ell-1} C_1^{-m} \sum_{\hat{J}:\text{length of }\hat{J}=m} \gamma_+(E_{\hat{J}})$$
$$+ A\, C_1^{-\ell} \sum_{J:\text{length of }J=\ell,\, J\text{ is not stopping}} \gamma_+(E_J)\,.$$

Let now (3.14) come into play again. Then (this is an entire analog of (6.6))

(6.12)
$$\sum_{m=0}^{\ell-1} C_1^{-m} \sum_{\hat{J}:\text{length of }\hat{J}=m} \gamma_+(E_{\hat{J}})$$
$$+ C_1^{-\ell} \sum_{J:\text{length of }J=\ell,\, J\text{ is not stopping}} \gamma_+(E_J) \le \gamma_+(E)\,.$$

Combine (6.12) and (6.11) to obtain finally the final inequality (6.7):
$$\gamma(E) \le A\,\gamma_+(E)\,.$$

We are left to justify (6.4). The consideration is based on the largeness of ℓ and it is as follows. Our final goal is to prove the last inequality $\gamma(E) \le A\,\gamma_+(E)$. But it is enough to obtain it only for E's which are a finite union of closed cubes with sides parallel to coordinate planes. Notice that after a large number ℓ of splitting steps we get the intersection of such an E with cubes of very small diameter (use (3.15)). Such an intersection $\mathcal{E}$ consists of at most L (here L depends only on the dimension n) parallelepipeds. The surface $\partial\mathcal{E}$ of this collection of parallelepipeds is formed by finitely many (say M, M depends only on n) parallelepipeds of dimension $n-1$. Let us choose the one which has the largest $\mathcal{H}^{n-1}$ measure. Call it $\mathcal{F}$. Then
$$\gamma(\mathcal{E}) \le \mathcal{H}^{n-1}(\partial\mathcal{E}) \le M\,\mathcal{H}^{n-1}(\mathcal{F})\,.$$

But it is obvious that for any parallelepiped of dimension $n-1$
$$\mathcal{H}^{n-1}(\mathcal{F}) \le A\,\gamma_+(\mathcal{F})\,.$$

Here A depends only on n. Combining the last two inequalities, one gets
$$\gamma(\mathcal{E}) \le A\,M\,\gamma_+(\mathcal{E})\,.$$

This is what gives (6.4) for sets E consisting of a finite union of closed cubes with sides parallel to coordinate planes. And this is the only situation we need.

We are left to prove (6.9) or, equivalently, to prove Theorem 6.1.

CHAPTER 7

The First Reduction to Nonhomogeneous Tb Theorem

We have already prepared lots of tools for this reduction. We start with a compact E that supports a distribution S with $\Phi * S \in L(E, 1)$. Of course we choose S such that $\langle 1, S \rangle = \gamma(E)$, which is obviously possible. Now let us construct measures μ and ν exactly as in Chapter 4. These are measures on

$$F := \cup_{i=1}^{N} Q_i \,.$$

The reader should take a look at Section 3.5 to see that the constant C_2 was introduced in (3.27). Recall that the ball $B(x, R), x \in F$ is called C_2 non-Ahlfors if $\mu(B(x, r)) > C_2 R^{n-1}$. Denote by $R(x)$ the supremum of radii of all non-Ahlfors balls centered at x. Then $R(x)$ is called the C_2 Ahlfors radius at point x.

Measures μ, ν have the following properties:

$$(7.1) \qquad |\nu(R)| \le A\, \ell(R)^{n-1} \;\; \forall \text{ cube } R \in \mathbb{R}^n \,,$$

$$(7.2) \qquad \int_{F \setminus H} V_*^{\nu}(x)\, d\mu(x) \le A\, \mu(F) \,,$$

where $H = \cup_k B(x_k, 3R(x_k))$ was introduced in Section 3.5. It contains all non-Ahlfors balls. The constant C_2 from (3.27) allows us to control the size of H because (see (3.29))

$$(7.3) \qquad \sum_k R(x_k)^{n-1} \le \frac{1}{C_2} \mu(F) \,.$$

At the same time H is sufficiently large in the sense that all non-Ahlfors balls are contained in H, and moreover

$$(7.4) \qquad \operatorname{dist}(x, F \setminus H) \ge R(x), \;\; \forall x \in F \,.$$

Here is our first reduction.

THEOREM 7.1. *Let finite numbers $\hat{A}, A^*, C_1, A_c$ be given. Let a positive measure μ supported on F and complex measure $\nu = b\, d\mu$ satisfy the following assumptions:*

$$(7.5) \qquad \|b\|_{\infty} \le \hat{A} \,,$$

$$(7.6) \qquad |\nu(F)| \ge C_1^{-1} \mu(F) \,,$$

$$(7.7) \qquad \int_{F \setminus H} V_*^{\nu}(x)\, d\mu(x) \le A^*\, \mu(F) \,,$$

$$(7.8) \qquad |\nu(R)| \le A_c\, \ell(R)^{n-1} \;\; \forall \text{ cube } R \in \mathbb{R}^n \,.$$

Then there exist constants $q \in (0,1)$, $C_3 < \infty$, $C_4 < \infty$, *and a closed subset* F_0 *of* F *such that* q, C_3, *and* C_4 *depend only on* n, $\hat{A}$, A^*, C_1, A_c *and*

$$(7.9) \qquad \mu(F_0) \geq (1-q)\mu(F) \,,$$

$$(7.10) \qquad \mu(B(x,r)) \leq C_3\, r^{n-1} \ \forall x \in F_0 \,,$$

$$(7.11) \qquad \|V^\mu : L^2(F_0, \mu) \to L^2(F_0, \mu)\| \leq C_4 \,.$$

Remarks. 1) The kernel Ψ of V is antisymmetric, so the boundedness of operator V^μ can be understood in the sense of the following bound on its symmetrized bilinear form (see for example [**47**]):

$$(7.12)$$
$$|\int \Psi(x-y)(\varphi(x)\psi(y) - \varphi(y)\psi(x))\, d\mu(x)\, d\mu(y)| \leq C_4\|\varphi\|_{L^2(F_0,\mu)}\|\psi\|_{L^2(F_0,\mu)} \,,$$

for any pair of C^∞ functions with compact support.
2) It is not too difficult to see that (7.10) follows from (7.11). But we prefer to keep both claims in the statement of Theorem 7.1 because we will start by proving (7.10).

Let us prove Theorem 6.1 provided that we know Theorem 7.1.

Proof. Let $F := \cup_{i=1}^N Q_i$, μ, ν constructed in Chapter 4. It is obvious from this construction, from $\langle 1, S_0 \rangle = \gamma(E)$, and from $\sum_{i=1}^N g_i = 1$ on E_0 that

$$(7.13) \qquad \nu(F) = \gamma(E) \,.$$

Also $\mu(F) = \sum_{i=1}^N \gamma(2Q_i \cap E)$. Then (6.2) of Theorem 6.1 ensures (7.6). We already saw in Chapter 4 that our measures satisfy (7.5), (7.7), and (7.8). If we apply Theorem 7.1 we get measure $\mu|F_0$, which we can use to estimate $\gamma_+(F)$:

$$\gamma_+(F) \geq C_4^{-1}\mu(F_0) \geq (1-q)C_4^{-1}\mu(F)$$
$$\geq (1-q)C_4^{-1}\hat{A}^{-1}|\nu(F)| = (1-q)C_4^{-1}\hat{A}^{-1}\gamma(E) \,.$$

The last equality is (7.13). On the other hand (3.13) implies $\gamma_+(F) \leq C_0\gamma_+(E)$. Combine the last two inequalities to get

$$\gamma(E) \leq A\gamma_+(E)$$

with $A = (1-q)^{-1}C_0C_1C_4\hat{A}$. Theorem 6.1 is reduced to Theorem 7.1. $\qquad\square$

We will reduce Theorem 7.1 to another result (second reduction). The reduction takes several steps.

The first step is to prove the existence of $q_1 \in (0,1)$ and $C_3 < \infty$ such that they depend only on $\hat{A}, C_1, A_c$ and such that there exists $F_1 \subset F$, $\mu(F_1) \geq (1-q_1)\mu(F)$ satisfying (7.10). There are many ways to prove this. We prefer the way, which will serve as the model for future reductions. Let $\mathcal{D}$ be a dyadic lattice of cubes in $\mathbb{R}^n$. We consider cubes R such that (we need again $H = \cup_k B(x_k, 3R(x_k))$ introduced in Section 3.5; see also (7.3)):

$$(7.14) \qquad \exists k \,:\, R \cap B(x_k, 3R(x_k)) \neq \emptyset, \ \ 10R(x_k) \leq \ell(R) < 20R(x_k) \,.$$

Among such R's we choose the maximal ones, call them $\{R_j\}_{j \in I_H}$. Let

$$(7.15) \qquad H_{\mathcal{D}} := \cup_{j \in I_H} R_j \,.$$

It is an open set and we want to show that we can choose F_1 to be $F \setminus H_{\mathcal{D}}$. Let us write the chain of inequalities. The constant C_2 from (7.3) will be involved. Recall that C_2 was introduced in Section 3.5, and constant 100 was involved in its definition. We will need to augment C_2, so we use the same definition but replace 100 by some huge constant > 100 which depends on $C_1, A_c, \hat{A}$. We will see now how to modify C_2 after writing the following chain of inequalities:

$$C_1^{-1}\mu(F)$$
$$\leq |\nu(F)| \leq |\nu(F \setminus H_{\mathcal{D}})| + \sum_{j \in I_H} |\nu(R_j)| \leq \hat{A}\mu(F \setminus H_{\mathcal{D}}) + A_c \sum_{j \in I_H} \ell(R_j)^{n-1}$$
$$\leq \hat{A}\mu(F \setminus H_{\mathcal{D}}) + 20a(n)A_c \sum_k R(x_k)^{n-1}$$
$$\leq \hat{A}\mu(F \setminus H_{\mathcal{D}}) + \frac{20a(n)A_c}{C_2} \sum_k \mu(B(x_k, R_k))$$
$$\leq \hat{A}\mu(F \setminus H_{\mathcal{D}}) + \frac{20a(n)A_c}{C_2}\mu(F) \,.$$

The constant $a(n)$ denotes the maximal number of dyadic cubes of family I_H intersecting ball $B(x_k, R(x_k))$. By construction (see (7.14)) it depends only on the dimension. Constant 20 appears from the same (7.14). Constant C_2 in the denominator appeared from (7.3). Let us choose it to be equal to (look at (3.27) for $A_1(n), A_2(n)$)

(7.16) $\qquad C_3 = 40a(n)A_cC_1 + 100^{n-1}\, A_1(n)A_2(n)\, C_1 > 40a(n)A_cC_1 \,.$

Then

(7.17) $$\sum_{j \in I_H} |\nu(R_j)| \leq \frac{20a(n)A_c}{C_3}\mu(F) \,,$$

and

$$\hat{A}\mu(F \setminus H_{\mathcal{D}}) > \frac{1}{2}C_1^{-1}\mu(F) \,.$$

Let us choose $q_1 = 1 - \frac{1}{2}C_1^{-1}\hat{A}^{-1}$,

(7.18) $$F_1 = F \setminus H_{\mathcal{D}} \,.$$

Then

$$\mu(F_1) > (1 - q_1)\mu(F) \,,$$

and let us see that F_1 satisfies (7.10) (with F_1 instead of F_0 of course). Construction in Section 3.5 of course can be performed for any constant C_2. In particular, we know from (7.4) that all C_3 non-Ahlfors balls are contained in $H \subset H_{\mathcal{D}}$. Thus $F_1 = F \setminus H_{\mathcal{D}}$ centers only C_3 Ahlfors balls.

We start the second step of the proof of Theorem 7.1 by the following conventions. We assume that

(7.19) $$F \subset B(0, \frac{1}{10}), \quad F \subset \frac{1}{4}\mathbf{Q} \,,$$

where $\mathbf{Q}$ is a unit cube of dyadic lattice $\mathcal{D}$. In the future $\mathcal{D}$ will vary, but we always assume property (7.19).

Notice that we have $d\nu = b\,d\mu$ and by (7.6) and (7.19)

$$(7.20) \qquad |\nu(\mathbf{Q})| = |\int_{\mathbf{Q}} b\,d\mu| > C_1^{-1}\mu(\mathbf{Q})\,.$$

Let us fix $C_5 \geq C_1$ (we will choose it later depending on C_1) and let us call $Q \in \mathcal{D}$ the C_5^{-1} accretive cube if

$$(7.21) \qquad |\nu(Q)| = |\int_Q b\,d\mu| > C_5^{-1}\mu(Q)\,.$$

For example, $\mathbf{Q}$ is C_5^{-1} accretive. We call dyadic subcubes of $\mathbf{Q}$ (from the same lattice) C_5 non-accretive if the opposite inequality holds.

The second step will consist in finding $q_2 \in (0,1), q_2 > q_1$ and $F_2 \subset F_1$, $\mu(F_2) > (1-q_2)\mu(F)$ such that all C_5^{-1} non-accretive cubes are disjoint with F_2. Let $\{R_j\}_{j \in I_T}$ be maximal C_5^{-1} non-accretive subcubes. All C_5^{-1} non-accretive subcubes (in lattice $\mathcal{D}$) are inside the set

$$(7.22) \qquad T_{\mathcal{D}} := \cup_{j \in I_T} R_j\,.$$

It is an open set and we want to show that we can choose F_2 to be $F \backslash (H_{\mathcal{D}} \cup T_{\mathcal{D}})$. Among the cubes of $H_{\mathcal{D}}$ and $T_{\mathcal{D}}$ choose the maximal ones, call them $\{R_j\}_{j \in I_{HT}}$.

Let the collection of indices $\{I_{HT}\}$ be such that $H_{\mathcal{D}} \cup T_{\mathcal{D}} = \cup_{j \in I_{HT}} R_j$. We will see now how to choose C_5 after writing the following chain of inequalities (we use (7.17) in it):

$$C_1^{-1}\mu(F) \leq |\nu(F)| \leq |\nu(F \backslash (H_{\mathcal{D}} \cup T_{\mathcal{D}})| + \sum_{j \in I_{HT}} |\nu(R_j)|$$

$$\leq |\nu(F \backslash (H_{\mathcal{D}} \cup T_{\mathcal{D}})| + \sum_{j \in I_H} |\nu(R_j)| + \sum_{j \in I_T} |\nu(R_j)|$$

$$\leq \hat{A}\mu(F \backslash H_{\mathcal{D}}) + \frac{20a(n)A_c}{C_3}\mu(F) + C_5^{-1}\sum_{j \in I_T}\mu(R_j)$$

$$\leq \hat{A}\mu(F \backslash H_{\mathcal{D}}) + \frac{20a(n)A_c}{C_3}\mu(F) + C_5^{-1}\mu(F)$$

$$\leq \hat{A}\mu(F \backslash H_{\mathcal{D}}) + (\frac{20a(n)A_c}{C_3} + C_5^{-1})\mu(F)\,.$$

Constant C_3 was fixed in (7.16). Let us choose $C_5 = 4C_1$. Then the last chain of inequalities give

$$\hat{A}\mu(F \backslash (H_{\mathcal{D}} \cup T_{\mathcal{D}})) > \frac{1}{4}C_1^{-1}\mu(F)\,.$$

Let us choose $q_2 = 1 - \frac{1}{4}C_1^{-1}\hat{A}^{-1}$,

$$(7.23) \qquad F_2 = F \backslash (H_{\mathcal{D}} \cup T_{\mathcal{D}})\,.$$

Then

$$\mu(F_2) > (1-q_2)\mu(F)\,,$$

and all C_5^{-1} non-accretive subcubes are disjoint with F_2, and F_2 centers only C_3 Ahlfors discs.

The third step consists in finding $q_3 \in (0,1), q_3 > q_2 > q_1$ and $F_3 \subset F_2$, $\mu(F_3) > (1-q_3)\mu(F)$ such that all C_5^{-1} non-accretive cubes are disjoint with F_3, F_3 centers only C_3 Ahlfors balls, and V_*^ν is uniformly bounded on F_3 by constant

C_6, which will depend only on $A^*, A_c, \hat{A}, C_1$. We start with (7.7) (it has not been used yet). We set

$$S_0 = S_0(s) = \{x \in F : V_*^{\nu}(x) > s\},$$

where s is some large constant to be chosen later. Then for $x \in S_0$, let

$$\varepsilon(x) = \sup\{\varepsilon : \varepsilon > 0, |V_{\varepsilon}^{\nu}(x)| > s\}.$$

Otherwise, we set $\varepsilon(x) = 0$. Let

(7.24) $$S = F \cap \cup_{x \in F} B(x, \varepsilon(x)).$$

We want to show that $\mu(S)$ is small if s is large. For $S_0(s)$ this would be clear by (7.7). So let us show that $S \subset S_0(s/2)$ if s is sufficiently large.

LEMMA 7.2. *If $y \in S \setminus H_{\mathcal{D}} = S \cap F_1$, then*

(7.25) $$V_*^{\nu}(y) > s - A\hat{A}C_3.$$

Proof. For such a y we have $x \in F$ and $\varepsilon_0 > 0$ such that $|V_{\varepsilon_0}^{\nu}(x)| > s$, $|y - x| < \varepsilon_0$. We will show that

(7.26) $$|V_{\varepsilon_0}^{\nu}(y) - V_{\varepsilon_0}^{\nu}(x)| \leq A\hat{A}C_3.$$

Then (7.25) follows. We have ($B_0 := B(y, \varepsilon_0)$)

$$|V_{\varepsilon_0}^{\nu}(y) - V_{\varepsilon_0}^{\nu}(x)| \leq |V_{\varepsilon_0}^{\nu|2B_0}(y)| + |V_{\varepsilon_0}^{\nu|2B_0}(x)| +$$
$$|V_{\varepsilon_0}^{\nu|(F \setminus 2B_0)}(y) - V_{\varepsilon_0}^{\nu|(F \setminus 2B_0)}(x)| =: I + II + III.$$

Notice that $|I| + |II| \leq A\frac{|\nu|(2B_0)}{\varepsilon_0^{n-1}}$ by the property of the kernel Ψ of V: $|\Psi(x)| \leq A/|x|^{n-1}$. Then

$$|I| + |II| \leq A\hat{A}\frac{\mu(2B_0)}{\varepsilon_0^{n-1}} \leq A\hat{A}C_3.$$

The last inequality uses the fact that $y \in S \setminus H_{\mathcal{D}} = S \cap F_1$, and so the ball centered at $y \in F_1$ is C_3 Ahlfors by the property of F_1. To estimate III, write

$$|III| \leq \hat{A}\int_{F \setminus 2B_0} \frac{A\varepsilon_0}{|z - y|^n} d\mu(z)$$

by the property of the kernel Ψ of V: $|\Psi(z - x) - \Psi(z - y)| \leq \frac{A|x-y|}{|z-x|^n}$ if $|x - y| \leq \frac{1}{2}|z - y|$. Now the estimate is standard. We denote by $A_k = \{z : 2^k\varepsilon_0 \leq |z - y| \leq 2^{k+1}\varepsilon_0\}$, $k = 1, 2, \ldots$. Then

$$|III| \leq \hat{A}\sum_k \frac{A\varepsilon_0\mu(A_k)}{2^{kn}\varepsilon_0^n} \leq A\hat{A}C_3\sum_k \frac{1}{2^k}.$$

Lemma 7.2 is completely proved. $\qquad\square$

Now let us choose $s = \max(2A\hat{A}C_3, 2A^*(1 - q_2)^{-1})$. Then of course

$$\mu(S) \leq \mu(S_0(s/2)) \leq \frac{2}{s}\int V_*^{\nu}(x) d\mu(x) \leq \frac{2}{s}A^* \leq \frac{1}{2}(1 - q_2).$$

Let us set $F_3 := F_2 \setminus S$. We have just proved that $\mu(F_3) \geq (1 - q_3)\mu(F)$, where $1 - q_3 = \frac{1}{2}(1 - q_2)$. For F_3: 1) all balls centered at F_3 are C_3 Ahlfors, 2) all C_5^{-1} non-accretive cubes of $\mathcal{D}$ are disjoint with F_3, and 3) $V_*^{\nu}(x) \leq C_6$ on F_3, where $C_6 = \max(2A\hat{A}C_3, 2A^*(1 - q_2)^{-1})$.

We have reduced Theorem 6.1 to Theorem 7.1. Now we are ready to make the second reduction. We reduce Theorem 7.1 to Theorem 8.1.

CHAPTER 8

The Second Reduction

We use the constants $\hat{A}, q_i, C_i$ from the previous section. We can use notation $H_{\mathcal{D}}, T_{\mathcal{D}}$, and S, F_1, F_2, F_3 from the previous section.

THEOREM 8.1. *Let positive measure μ and complex measure $\nu = b\,d\mu$ be supported on F satisfying (7.19). Suppose that there exist constants $q \in (0,1), C_3 < \infty, C_5 < \infty, C_6 < \infty$ such that for any dyadic lattice $\mathcal{D}$ satisfying (7.19), there exists a compact subset $F_{\mathcal{D}}$ of F such that $\mu(F_{\mathcal{D}}) \geq (1-q)\mu(F)$ and such that*

1) all balls centered at $F_{\mathcal{D}}$ are C_3 Ahlfors,

2) all C_5^{-1} non-accretive cubes of $\mathcal{D}$ are disjoint with $F_{\mathcal{D}}$, and $V_*^{\nu}(x) \leq C_6$ on $F_{\mathcal{D}}$.

Suppose also that (7.5) holds with constant $\hat{A}$.

Then there exist constants $q_0 \in (0,1), C_4 < \infty$, depending on $q, C_3, C_5, C_6, \hat{A}$, and a compact subset F_0 of F such that

$$\mu(F_0) \geq (1-q_0)\mu(F)\,, \tag{8.1}$$

$$\mu(B(x,r)) \leq C_3\,r^{n-1}\ \forall x \in F_0\,, \tag{8.2}$$

$$\|V^{\mu} : L^2(F_0, \mu) \to L^2(F_0, \mu)\| \leq C_4\,. \tag{8.3}$$

We can now prove Theorem 7.1 using Theorem 8.1.

Proof. In Chapter 7 we constructed q_3 and F_3 for every $\mathcal{D}$ satisfying (7.19). Set $q = q_3, F_{\mathcal{D}} = F_3$, and apply Theorem 8.1. Then we have all the statements of Theorem 7.1. $\qquad\square$

The proof of Theorem 8.1 will be long and difficult. Approximately 70 subsequent pages are devoted to the proof. It will be a reduction to yet another theorem. We start the proof by introducing suppressed kernels, invented by F. Nazarov.

8.1. Suppressed kernels

Let θ stand for a nonnegative Lipschitz function in $\mathbb{R}^n$:

$$\theta \geq 0\,, \quad |\theta(x) - \theta(y)| \leq |x - y|\,. \tag{8.4}$$

Certain kernels were very instrumental in [**47**] for proving fancy Tb theorems (this is what we are going to do next). Then in Tolsa's work these kernels played a very important part because a fancy Tb theorem is very essential in his work. Here are those kernels:

$$k_{\theta}(z, \zeta) := \frac{z - \zeta}{|z - \zeta|^2 + \theta(z)\theta(\zeta)}\,, \quad z, \zeta \in \mathbb{C}\,. \tag{8.5}$$

This is a Cauchy kernel suppressed by θ. It is a Calderón–Zygmund kernel (see [**47**] or below).

We are dealing with a multidimensional version of the Cauchy kernel here. Namely, we are dealing with our kernel $\Psi = (\Psi^1, \ldots, \Psi^n)$, where

$$(8.6) \qquad \Psi^i(x, y) := \frac{x_i - y_i}{|x_i - y_i|^n}, \quad x, y \in \mathbb{R}^n .$$

The index i will be unessential for us, so let $k(x, y)$ denote *any* $\Psi^i(x, y)$. What is essential is its Calderón–Zygmund properties (below $m = n - 1$ but we prefer to use the letter m for the sake of generality):

$$(8.7) \qquad \begin{aligned} |k(x, y)| &\leq \frac{c_1}{|x - y|^m} \, ; |k(x', y) - k(x, y)| \\ &\leq \frac{c_2|x - x'|^\tau}{|x - y|^{m+\tau}} \, , \ \frac{|x - x'|}{|x - y|} \leq \frac{1}{2} \, , \ x, x', y \in \mathbb{R}^n . \end{aligned}$$

We require also the symmetric condition

$$|k(x, y') - k(x, y)| \leq \frac{c_2|y' - y|^\tau}{|x - y|^{m+\tau}} \, , \ \frac{|y - y'|}{|x - y|} \leq \frac{1}{2} \, , \ x, y', y \in \mathbb{R}^n .$$

Constants $c_{CZ} := (m > 0, \ \tau \in (0, 1], \ c_1 < \infty, \ c_2 < \infty)$ are called Calderón–Zygmund parameters of the kernel. In our case $m = n - 1, \tau = 1$, and c_1, c_2 depend only on n. Notice also that in our case

$$k(x, y) = -k(y, x) .$$

Such kernels are called *antisymmetric* Calderón–Zygmund kernels. All our kernels will be antisymmetric Calderón–Zygmund kernels.

The smallest number m in (8.7) is called *the order* of the kernel k. Now we want to suppress any real-valued, antisymmetric Calderón–Zygmund kernel of order m by means of nonnegative Lipschitz function θ on $\mathbb{R}^n$. So let θ satisfy (8.4). We set

$$(8.8) \qquad k_\theta(x, y) := k(x, y) \cdot \frac{1}{1 + k^2(x, y)\theta(x)^m\theta(y)^m} .$$

LEMMA 8.2. *Let k be any real-valued, antisymmetric Calderón–Zygmund kernel of order m, and let θ satisfy (8.4). Then k_θ is an antisymmetric Calderón–Zygmund kernel of order m, $k_\theta(x, y) = k(x, y)$ if $\theta(x) = 0$ or $\theta(y) = 0$, and*

$$(8.9) \qquad |k_\theta(x, y)| \leq C(m) \min\left[\frac{1}{\theta(x)^m}, \frac{1}{\theta(y)^m}\right] .$$

Proof. Everything is obvious except (8.9) and (8.7) for k_θ. Let us first prove (8.9). We have symmetry with respect to x, y. Let $\theta(x) = \max(\theta(x), \theta(y))$.

$$(8.10) \qquad |k_\theta(x, y)| \leq \frac{|k(x, y)|^{-1}}{k(x, y)^{-2} + \theta(x)^m\theta(y)^m} \leq \frac{|k(x, y)|^{-1}}{\theta(x)^m\theta(y)^m} .$$

In fact, (8.7) ensures that $|x - y|^m \leq c_1|k(x, y)|^{-1}$. Suppose first that $c_1|k(x, y)|^{-1} \leq \frac{\theta(x)^m}{2^m}$. Then we can continue by $|x - y|^m \leq c_1|k(x, y)|^{-1} \leq \frac{\theta(x)^m}{2^m}$. Therefore, $|x - y| \leq \frac{\theta(x)}{2}$. But $\theta(y) \geq \theta(x) - |x - y|$, and we get $\theta(y) \geq \frac{\theta(x)}{2}$. Combining (8.10) with our assumption $c_1|k(x, y)|^{-1} \leq \frac{\theta(x)^m}{2^m}$ we get $|k_\theta(x, y)| \leq A\frac{1}{\theta(x)^m}$.

Now suppose that $c_1|k(x, y)|^{-1} > \frac{\theta(x)^m}{2^m}$. Then $|k_\theta(x, y)| \leq |k(x, y)| \leq c_1\frac{2^m}{\theta(x)^m}$. So in both cases (8.9) is proved.

We want to prove (8.7) for k_θ. Let $|x - x'| \leq \frac{1}{2}|x - y|$.

We first assume that $\theta(x') \leq \theta(x)$. Then we estimate $|k_\theta(x, y) - k_\theta(x', y)|$ as follows:

$$|k_\theta(x, y) - k_\theta(x', y)| \leq |k(x, y) - k(x', y)| \cdot \frac{1}{1 + k^2(x, y)\theta(x)^m\theta(y)^m}$$

$$+ |k(x', y)| \cdot \frac{|k^2(x, y)\theta(x)^m - k^2(x', y)\theta(x')^m|\theta(y)^m}{(1 + \ldots)(1 + \ldots)}$$

$$\leq \frac{c_2|x - x'|^\tau}{|x - y|^{m+\tau}} + |k(x', y)| \cdot \frac{|k^2(x, y) - k^2(x', y)|\theta(x')^m\theta(y)^m}{(1 + \ldots)(1 + \ldots)}$$

$$+ |k(x', y)| \cdot \frac{k^2(x, y)[\theta(x)^m - \theta(x')^m]\theta(y)^m}{(1 + \ldots)(1 + \ldots)} =: I + II + III.$$

Term I is already in the form we want. As to II,

$$II \leq |k(x, y) - k(x', y)| \cdot IV,$$

where

$$IV := \frac{|k(x', y)|(|k(x, y)| + |k(x', y)|)\theta(x')^m\theta(y)^m}{(1 + \ldots)(1 + \ldots)}.$$

It is sufficient, of course, just to bound IV by an absolute constant. It is bounded by

$$\frac{k^2(x', y)\theta(x')^m\theta(y)^m}{(1 + k^2(x', y)\theta(x')^m\theta(y)^m)(1 + k^2(x, y)\theta(x)^m\theta(y)^m)}$$

$$+ \frac{|k(x, y)||k(x', y)|\theta(x')^m\theta(y)^m}{(1 + \ldots)(1 + \ldots)}.$$

The first term is obviously bounded by 1. If $|k(x, y)| \leq |k(x', y)|$, then the second term becomes the first one. Otherwise the second term is bounded by

$$\frac{k^2(x, y)\theta(x')^m\theta(y)^m}{(1 + k^2(x', y)\theta(x')^m\theta(y)^m)(1 + k^2(x, y)\theta(x)^m\theta(y)^m)}.$$

But we assumed that $\theta(x') \leq \theta(x)$, so this expression is also bounded by 1.

Let us consider now term III.

$$III \leq |x - x'||k(x', y)|\frac{m\theta(x)^{m-1}\theta(y)^m k^2(x, y)}{(1 + \ldots)(1 + \ldots)}.$$

We consider two cases: 1) $\theta(x) \leq |x - y|$, and 2) $|x - y| < \theta(x)$. If the first case occurs, then $\theta(y) \leq 2|x - y|$ and

$$III \leq \frac{Ac_2}{|x - y|^{3m}}|x - y|^{2m-1}|x - x'| \leq \frac{Ac_2|x - x'|}{|x - y|^{m+1}} \leq \frac{Ac_2|x - x'|^\tau}{|x - y|^{m+\tau}},$$

where $\tau \in (0, 1]$ is from (8.7) for k. Finally $|k_\theta(x, y) - k_\theta(x', y)| \leq \frac{c|x-x'|^\tau}{|x-y|^{m+\tau}}$. (In the last inequality we used $|x - x'| \leq \frac{1}{2}|x - y|$.)

If the second case occurs, then

$$III \leq |k(x', y)|\frac{|x - x'|}{|x - y|}\frac{k^2(x, y)\theta(x)^m\theta(y)^m}{(1 + k^2(x, y)\theta(x)^m\theta(y)^m)(1 + \ldots)} \leq$$

$$\frac{c_1|x - x'|}{|x - y|^{m+1}} \leq \frac{Ac_1|x - x'|^\tau}{|x - y|^{m+\tau}},$$

where $\tau \in (0,1]$ is from (8.7) for k. (In the last inequality we used again $|x - x'| \leq \frac{1}{2}|x - y|$.)

If $\theta(x) \leq \theta(x')$, we use the symmetric considerations with x and x' interchanging their parts. $\square$

Let us introduce one more (and more standard) regularization of the kernel k of order m (from(8.7)) by function θ (with (8.4)):

$$(8.11) \qquad c_\theta(x,y) = k(x,y), \text{ if } |y - x| > \theta(x); \ \ c_\theta(x,y) = 0, \text{ otherwise}.$$

Definition. Let positive measure μ be given. The integral operator with kernel k_θ with respect to μ will be called T^θ, and the operator with kernel c_θ will be called C^θ. This definition correctly defines the operators if k is a bounded kernel (for example if θ is bounded away from zero). In other cases our kernels are always antisymmetric. The operator with antisymmetric Calderón–Zygmund kernel can be always introduced as a *canonical value*: we understand $T^\theta f = T^\theta_\mu f$ for smooth f in the sense that for any smooth test function g,

$$\langle T^\theta f, g \rangle = \frac{1}{2} \int \int k_\theta(x,y)[f(y)g(x) - f(x)g(y)] \, d\mu(x)d\mu(y).$$

We also introduce the following maximal functions:

$$M_{m,\theta}f(x) := \sup_{r > \theta(x)} \frac{1}{r^m} \int_{B(x,r)} |f(y)| \, d\mu(y),$$

$$M_{\mu,\theta}f(x) := \sup_{r > \theta(x)} \frac{1}{\mu(B(x,3r))} \int_{B(x,r)} |f(y)| \, d\mu(y),$$

$$M_\mu f(x) := \sup_{r > 0} \frac{1}{\mu(B(x,3r))} \int_{B(x,r)} |f(y)| \, d\mu(y).$$

Fix a $B < \infty$. Let $R(x)$ denote the supremum of radii of all B non-Ahlfors balls centered at x.

LEMMA 8.3. *Let real antisymmetric k satisfy (8.7). Let θ satisfy (8.4). Then*

$$(8.12) \qquad |(T^\theta)f(x) - (C^\theta)f(x)| \leq A \, M_{m,\theta}f(x).$$

If $\theta(x) \geq R(x)$, then

$$(8.13) \qquad |(T^\theta)f(x) - (C^\theta)f(x)| \leq AB \, M_{\mu,\theta}f(x) \leq AB \, M_\mu f(x).$$

In particular,

$$(8.14) \qquad \|(T^\theta)f(x) - (C^\theta)f(x)\|_{L^2(\mu)} \leq AB\|f\|_{L^2(\mu)}.$$

Proof. It is obvious that

$$(8.15) \qquad M_{m,\theta}f(x) \leq AB \, M_{\mu,\theta}f(x)$$

if $\theta(x) \geq R(x)$. It is sufficient to prove (8.12), as (8.13) follows immediately from (8.15), and (8.14) is just the consequence of a well-known fact that M_μ is a bounded operator in $L^2(\mu)$.

To prove (8.12) we fix $x \in \mathbb{R}^n$ and estimate $k_\theta(x,y) - c_\theta(x,y)$. If $y \in B(x, \theta(x))$, then this difference is at most $\frac{C}{\theta(x)^m}$ (see (8.9)). If y is such that $|y - x| \geq \theta(x)$, then $k_\theta(x,y) - c_\theta(x,y)$ is equal to

$$(8.16) \quad |k_\theta(x,y) - k(x,y)| \leq |k_\theta(x,y)|\theta(x)^m \theta(y)^m k^2(x,y) \leq C\theta(x)^m \frac{1}{|x-y|^{2m}}.$$

We used (8.9) again. We also used (8.7) for k. Henceforth,

$$|(T^\theta)f(x) - (C^\theta)f(x)| \le \frac{C}{\theta(x)^m} \int_{B(x,\theta(x))} |f(y)| \, d\mu(y) +$$

$$C \int_{y \in \mathbb{R}^n \setminus B(x,\theta(x))} \frac{\theta(x)^m}{|x-y|^{2m}} |f(y)| \, d\mu(y) \,.$$

We can finish the proof in a standard way by splitting the integral over $\mathbb{R}^n \setminus B(x,\theta(x))$ into integrals over annuli $A_k := \{y : 2^k\theta(x) \le |y-x| < 2^{k+1}\theta(x)\}$ and repeating the standard argument as at the end of the proof of Lemma 7.2. Lemma 8.3 is completely proved. $\qquad\square$

Definition. Let μ be a positive measure, and k be a real-valued antisymmetric Calderón–Zygmund kernel, $\varepsilon > 0$, and k_θ, c_θ as above. Define

(8.17)
$$T_\varepsilon^\theta f(x) := \int_{y:|y-x|>\varepsilon} k_\theta(x,y) f(y) \, d\mu(y) \,;$$

$$T_*^\theta f(x) := \sup_\varepsilon |\int_{y:|y-x|>\varepsilon} k_\theta(x,y) f(y) \, d\mu(y)| \,.$$

(8.18)
$$C_\varepsilon f(x) := \int_{y:|y-x|>\varepsilon} k(x,y) f(y) \, d\mu(y) \,;$$

$$C_* f(x) := \sup_\varepsilon |\int_{y:|y-x|>\varepsilon} k(x,y) f(y) \, d\mu(y)| \,.$$

Remark. The estimates in Lemma 8.3 were based only on the estimate of the absolute value of the difference of the kernels C_θ, k_θ. This gives us for any $\varepsilon > 0$

(8.19)
$$|(T_\varepsilon^\theta)f(x) - (C_\varepsilon^\theta)f(x)| \le A \, M_{m,\theta} f(x) \,.$$

If $\theta(x) \ge R(x)$, then

(8.20)
$$|(T_\varepsilon^\theta)f(x) - (C_\varepsilon^\theta)f(x)| \le AB \, M_{\mu,\theta} f(x) \le AB \, M_\mu f(x) \,.$$

In particular,

(8.21)
$$\|(T_\varepsilon^\theta)f(x) - (C_\varepsilon^\theta)f(x)\|_{L^2(\mu)} \le AB\|f\|_{L^2(\mu)} \,.$$

In its turn, it is easy to see that these inequalities imply

(8.22)
$$|(T_*^\theta)f(x) - (C_*^\theta)f(x)| \le A \, M_{n-1,\theta} f(x) \,.$$

If $\theta(x) \ge R(x)$, then

(8.23)
$$|(T_*^\theta)f(x) - (C_*^\theta)f(x)| \le AB \, M_{\mu,\theta} f(x) \le AB \, M_\mu f(x) \,.$$

In particular,

(8.24)
$$\|(T_*^\theta)f(x) - (C_*^\theta)f(x)\|_{L^2(\mu)} \le AB\|f\|_{L^2(\mu)} \,.$$

Fix $s > 0$. Let μ have a compact support F. We set

$$S_0 = S_0(s) = \{x \in F : T_* f(x) > s\} \,.$$

Then for $x \in S_0$, let

$$\varepsilon(x) = \sup\{\varepsilon : \varepsilon > 0, |T_\varepsilon f(x)| > s\} \,.$$

Otherwise, we set $\varepsilon(x) = 0$. Bellow balls are open.

LEMMA 8.4. *Let kernel k with Calderón–Zygmund constant C and function θ be as above, and let*

(8.25) $$\theta(x) \geq \max[\varepsilon(x), R(x)],$$

where $R(x)$ denotes the supremum of radii of all B non-Ahlfors balls centered at x. Then

(8.26) $$T_*^\theta f(x) \leq A(s + BCM_{\mu,\theta} f(x)), \ \forall x \in \mathbb{R}^n.$$

Proof. Let x be such that $\theta(x) = 0$. Then $k_\theta(x, y) = k(x, y)$ for all y, and $T_*^\theta f(x) = T_* f(x) \leq s$. Now let $\theta(x) > 0$. Let now $0 < \varepsilon$. Consider the case $\varepsilon \leq \theta(x)$. Using (8.9) we write

$$|T_\varepsilon^\theta f(x)| \leq C \frac{1}{\theta(x)^m} \int_{B(x,\theta(x))} |f(y)|\, d\mu(y) + |T_{\theta(x)}^\theta f(x) - T_{\theta(x)} f(x)| + |T_{\theta(x)} f(x)|.$$

But $\theta(x) \geq \varepsilon(x)$ and $\varepsilon(x)$ is the supremum of ε such that $|T_\varepsilon f(x)| > s$. Hence, $|T_{\theta(x)} f(x)| \leq s$. As to the second term, we can use (8.16) to write

$$|T_{\theta(x)}^\theta f(x) - T_{\theta(x)} f(x)| \leq C \int_{y:|y-x|>\theta(x)} \frac{\theta(x)^m}{|x-y|^{2m}} |f(y)|\, d\mu(y) \leq ABC\, M_{\mu,\theta} f(x).$$

The last inequality follows from the assumption $\theta(x) \geq R(x)$. Now we can continue our estimate:

$$|T_\varepsilon^\theta f(x)| \leq ABC\, M_{\mu,\theta} f(x) + s.$$

Consider now the case $\varepsilon > \theta(x)$. Then $\varepsilon > \varepsilon(x)$ and by definition of $\varepsilon(x)$ we have $|T_\varepsilon f(x)| \leq s$. Then

$$|T_\varepsilon^\theta f(x)| \leq |T_\varepsilon^\theta f(x) - T_\varepsilon f(x)| + |T_\varepsilon f(x)| \leq |T_\varepsilon^\theta f(x) - T_\varepsilon f(x)| + s.$$

But we have already estimated $|T_\varepsilon^\theta f(x) - T_\varepsilon f(x)| \leq C \int_{y:|y-x|>\varepsilon} \frac{\theta(x)^m}{|x-y|^{2m}} |f(y)|\, d\mu(y)$, which is bounded by $C\, M_{m,\theta} f(x)$ because $\varepsilon > \theta(x)$. But $\theta(x) \geq R(x)$ and so $M_{m,\theta} f(x) \leq B M_{\mu,\theta} f(x)$. Finally,

$$|T_\varepsilon^\theta f(x)| \leq BC\, M_{\mu,\theta} f(x) + s \leq BC\, M_{\mu,\theta} f(x) + s.$$

Lemma 8.4 is completely proved. $\qquad\square$

The next lemma is of the same nature. Notice that the set $K(s) := \{x \in \mathbb{R}^n : T_* f(x) \leq s\}$ is closed.

LEMMA 8.5. *Let kernel k with Calderón–Zygmund constant C and function θ be as above, and let*

(8.27) $$\theta(x) \geq \max[\mathrm{dist}(x, K(s)), R(x)],$$

where $R(x)$ denotes the supremum of radii of all B non-Ahlfors balls centered at x. Then

(8.28) $$T_*^\theta f(x) \leq A(s + BCM_{\mu,\theta} f(x)), \ \forall x \in \mathbb{R}^n.$$

Proof. Let $x : \theta(x) = 0$. Then $k_\theta(x, y) = k(x, y)$ for all y, and $T_*^\theta f(x) = T_* f(x) \leq s$. Now let $\theta(x) > 0$ and let z_0 be the closest to x point of $K(s)$. It lies outside of all B non-Ahlfors balls. Let $\varepsilon_0 = |z_0 - x|$. By assumption, $\theta(x) \geq \varepsilon_0$. Let $\varepsilon < \varepsilon_0$. Then

$$T_\varepsilon^\theta f(x) = \int_{\varepsilon < |y-x| \leq \varepsilon_0} k_\theta(x, y) f(y)\, d\mu(y) + T_{\varepsilon_0}^\theta f(x),$$

and by (8.9)

$$|T_\varepsilon^\theta f(x)| \leq \frac{C}{\theta(x)^m} \int_{|y-x| \leq \varepsilon_0} |f(y)|\, d\mu(y) + |T_{\varepsilon_0}^\theta f(x)|$$

$$\leq \frac{C}{\theta(x)^m} \int_{|y-x| \leq \theta(x)} |f(y)|\, d\mu(y) + |T_{\varepsilon_0}^\theta f(x)| \leq M_{\mu,\theta} f(x) + |T_{\varepsilon_0}^\theta f(x)|\,.$$

Thus, everything is reduced to the case when $\varepsilon \geq \varepsilon_0$. Recall that $\theta(x) \geq \varepsilon_0$ also.

The first case: $\varepsilon \geq \theta(x) \geq \varepsilon_0$. We can write

$$T_\varepsilon^\theta f(x) = (T_\varepsilon^\theta f(x) - T_{2\varepsilon}^\theta f(z_0)) + (T_{2\varepsilon}^\theta f(z_0) - T_{2\varepsilon} f(z_0)) + T_{2\varepsilon} f(z_0) =: I + II + III\,.$$

Clearly, $|III| \leq s$ as $z_0 \in K(s)$. Now Lemma 8.2 will give

$$|I| = \left| \int_{B(z_0, 2\varepsilon) \setminus \overline{B(x, \varepsilon)}} k_\theta(x, y) f(y)\, d\mu(y) \right|$$

$$\leq \frac{BC\, 15^m}{(15\varepsilon)^m} \int_{B(x, 3\varepsilon)} |f(y)\, d\mu(y) \leq BC\, M_{\mu,\theta} f(x)\,.$$

The last inequality holds because $\varepsilon \geq \theta(x) \geq R(x)$, and so $B(x, 15\varepsilon)$ is a B Ahlfors ball.

To estimate II we use (8.16):

$$|II| = |T_{2\varepsilon}^\theta f(z_0) - T_{2\varepsilon} f(z_0)| \leq C \int_{y:|y-z_0|>2\varepsilon} \frac{\theta(z_0)^m}{|y-z_0|^m} |f(y)|\, d\mu(y)\,.$$

But $\theta(z_0) \leq \theta(x) + |z_0 - x| \leq 2\varepsilon$. So, using the fact that $\varepsilon \geq \theta(x)$, we get

$$|II| \leq C \int_{y:|y-z_0|>2\varepsilon} \frac{2\varepsilon}{|y-z_0|^m} |f(y)|\, d\mu(y)$$

$$\leq C \int_{y:|y-x|>\varepsilon} \frac{\varepsilon}{|y-x|^m} |f(y)|\, d\mu(y) \leq AC\, M_{m,\theta} f(x)\,.$$

But $M_{m,\theta} f(x) \leq AB\, M_{\mu,\theta} f(x)$ by (8.15). The first case is done.

The second case: $\theta(x) > \varepsilon \geq \varepsilon_0$. The previous case shows that

$$|T_{\theta(x)}^\theta f(x)| \leq s + ABC\, M_{\mu,\theta} f(x)\,.$$

We are left to estimate

$$T_\varepsilon^\theta f(x) - T_{\theta(x)}^\theta f(x)\,.$$

It is equal to $\int_{\varepsilon < |y-x| \leq \theta(x)} k_\theta(x, y) f(y)\, d\mu(y)$. Using (8.9) once more we get

$$|T_\varepsilon^\theta f(x) - T_{\theta(x)}^\theta f(x)| \leq \frac{C}{\theta(x)^m} \int_{B(x, \theta(x))} |f(y)|\, d\mu(y) \leq AC\, M_{m,\theta} f(x)\,.$$

Now we use $\theta(x) \geq R(x)$ together with (8.15). And we are done. $\qquad \square$

8.2. From real-valued kernel to vector valued kernel

Definition. Let μ be a positive measure, and let k be an n-tuple $k = (k_1, \ldots, k_n)$ of real-valued antisymmetric Calderón–Zygmund kernels, $\varepsilon > 0$.

(8.29)
$$T_\varepsilon f(x) := \int_{y:|y-x|>\varepsilon} k(x, y) f(y)\, d\mu(y)\,;$$

$$T_* f(x) := \sup_\varepsilon \left| \int_{y:|y-x|>\varepsilon} k(x, y) f(y)\, d\mu(y) \right|\,.$$

Here $|\cdot|$ denote the Euclidean norm in $\mathbb{R}^n$.

Let θ be a Lipschitz function satisfying (8.4). We define $k_\theta := (k_{1,\theta}, \ldots, k_{n,\theta})$. In particular, we defined $T_\varepsilon^\theta f(x)$. We define $T_*^\theta f(x)$ as the sum of corresponding star operators for components.

It is easy to see that Lemma 8.4 and Lemma 8.5 hold true. We repeat.

Fix $s > 0$. Let μ have a compact support F. We set

$$S_0 = S_0(s) = \{x \in F : T_* f(x) > s\}.$$

Then for $x \in S_0$, let

$$\varepsilon(x) = \sup\{\varepsilon : \varepsilon > 0, |T_\varepsilon f(x)| > s\}.$$

Otherwise, we set $\varepsilon(x) = 0$.

LEMMA 8.6. *Let vector valued kernel* k *with Calderón–Zygmund constant* C *and function* θ *be as above, and let*

$$(8.30) \qquad \theta(x) \geq \max[\varepsilon(x), R(x)],$$

where $R(x)$ *denotes the supremum of radii of all* B *non-Ahlfors balls centered at* x. *Then*

$$(8.31) \qquad T_*^\theta f(x) \leq A(s + BCM_{\mu,\theta} f(x)), \ \forall x \in \mathbb{R}^n.$$

The set $K(s) := \{x \in \mathbb{R}^n : T_* f(x) \leq s\}$ is again closed.

LEMMA 8.7. *Let vector-valued kernel* k *with Calderón–Zygmund constant* C *and function* θ *be as above, and let*

$$(8.32) \qquad \theta(x) \geq \max[\operatorname{dist}(x, K(s)), R(x)],$$

where $R(x)$ *denotes the supremum of radii of all* B *non-Ahlfors balls centered ai* x. *Then*

$$(8.33) \qquad T_*^\theta f(x) \leq A(s + BCM_{\mu,\theta} f(x)), \ \forall x \in \mathbb{R}^n.$$

8.3. From one lattice to two lattices

We are coming back to the proof of Theorem 8.1. Let Ω denote the probability space of all lattices $\mathcal{D}$ satisfying (7.19). The probability measure on the space of such lattices can be introduced, for example, as normalized Lebesgue measure on all shifts of some fixed lattice, which are shifted with care, namely, preserving (7.19). If $\omega \in \Omega$, then $\mathcal{D}_\omega$ is a corresponding dyadic lattice satisfying (7.19).

The assumption of Theorem 8.1 says that for every $\omega \in \Omega$ there exists $F_\omega := F_{\mathcal{D}_\omega}$ such that several nice things happen on this set and $\mu(F_\omega) \geq \tau \mu(F)$, where $\tau = 1 - q > 0$. Sets F_ω wander over F, so their intersection can be empty, which means that there is no $x \in F$ that lies in all these sets (these sets are supposed to be "nice"). However, it is easy to see that there is a set of x's that has positive μ measure and consists of points, which lie in "nice" sets F_ω with positive fixed probability. We are under the assumptions of Theorem 8.1 in the next lemma.

LEMMA 8.8. *Let* $\tau > 0$. *Let for every* $\omega \in \Omega$ *be given a set* $F_\omega \subset F$ *such that* $\mu(F_\omega) \geq \tau \mu(F)$. *Then there exists a set* F_c *such that* $\mu(F_c) \geq \tau/2 \mu(F)$ *and such that* $P\{\omega \in \Omega : x \in F_\omega\} \geq \tau/2$ *for any* $x \in F_c$.

Proof. Let $G := \{(x, \omega) \in F \times \Omega : x \in F_\omega\}$. Set $p(x) := P\{\omega : x \in F_\omega\} = \int_\Omega \mathbf{1}_G(x, \omega) \, dP$. Then

$$\int_F p(x) \, d\mu(x) = \int_F \int_\Omega \mathbf{1}_G(x, \omega) \, dP \, d\mu$$

$$= \int_\Omega \int_F \mathbf{1}_G(x, \omega) \, d\mu \, dP = \int_\Omega \mu(F_\omega) \, dP \geq \tau \mu(F) \, .$$

Now we split $F = L \cup (F \setminus L)$, where $L := \{x \in F : p(x) < \tau/2\}$. Then the previous inequality gives

$$\tau \, \mu(F) \leq \int_L p(x) \, d\mu(x) + \int_{F \setminus L} p(x) \, d\mu(x) \leq (\tau/2) \, \mu(F) + \mu(F \setminus L) \, .$$

It remains to set $F_c = F \setminus L$, and the lemma is proved. $\qquad \square$

We also need $\Omega^2 = \{(\omega_1, \omega_2) : \omega_1, \omega_2 \in \Omega\}$. We use ω_1, ω_2 to denote two independent lattices $\mathcal{D}_1$, $\mathcal{D}_2$ from Ω. We need to see that there is a set of x's that has positive μ measure and consists of points, which lie in the intersection of "nice" sets F_{ω_1}, F_{ω_2} with positive fixed probability. We are under the assumptions of Theorem 8.1 in the next lemma.

LEMMA 8.9. *Let $\tau > 0$. Let for every $\omega \in \Omega$ we have a set $F_\omega \subset F$ such that $\mu(F_\omega) \geq \tau \, \mu(F)$. Then there exists a set $E \subset F$ such that $\mu(E) \geq (\tau/2) \, \mu(F)$ and such that $P\{\omega = (\omega_1, \omega_2) \in \Omega^2 : x \in F_{\omega_1} \cap F_{\omega_1}\} \geq (\tau/2)^2$ for any $x \in E$.*

Proof. Fix $x \in F$. Events $\{\omega = (\omega_1, \omega_2) \in \Omega^2 : x \in F_{\omega_1}\}$ and $\{\omega = (\omega_1, \omega_2) \in \Omega^2 : x \in F_{\omega_2}\}$ are independent. And probability of each of them is equal to $p(x)$ from Lemma 8.8. So $P\{x \in F_{\omega_1} \cap F_{\omega_2}\} = p(x)^2$. So we can choose $E = F_c$ with F_c from Lemma 8.8, and our lemma is proved. $\qquad \square$

8.4. Core suppression

Notation. Fix $\omega = (\omega_1, \omega_2) \in \Omega^2$, put $F_{(\omega_1, \omega_2)} = F_{\omega_1} \cap F_{\omega_2}$. Put $O(\omega_1, \omega_2) := B(0, 1/5) \setminus F_{(\omega_1, \omega_2)}$, $\theta_{(\omega_1, \omega_2)}(x) := \text{dist}(x, \mathbb{R}^n \setminus O_{(\omega_1, \omega_2)})$. From now on our probability space will be always Ω^2. So for brevity we will denote these objects $F_\omega, O_\omega, \theta_\omega$, meaning $\omega = (\omega_1, \omega_2)$.

Let us introduce *a core suppression function*. We keep in mind the assumptions and notation of Theorem 8.1. We use $\tau = 1 - q$, where q is from Theorem 8.1. We put

(8.34) $$\beta := \tau/2 = (1 - q)/2 \, ,$$

(8.35) $$\Theta_0(x) := \inf_{Z \subset \Omega^2, \, P\{Z\} \geq \beta^2} \sup_{\omega \in Z} \theta_\omega(x) \, .$$

For the sake of convenience we fix $\kappa > 0$ and also introduce

(8.36) $$\Theta(x) := \Theta_0(x) + \kappa \, .$$

CHAPTER 9

The Third Reduction

We are ready to formulate the third reduction of our main goal, namely, of Theorem 7.1. We use the notation of the second reduction (Theorem 8.1).

THEOREM 9.1. *Let positive measure μ and complex measure $\nu = b\,d\mu$ be supported on F satisfying (7.19). Suppose that there exist constants $q \in (0,1), C_3 < \infty, C_5 < \infty, C_6 < \infty$ such that for any dyadic lattice $\mathcal{D}$ satisfying (7.19) there exists a compact subset $F_{\mathcal{D}}$ of F such that $\mu(F_{\mathcal{D}}) \geq (1-q)\mu(F)$ and such that*
   1) all balls centered at $F_{\mathcal{D}}$ are C_3 Ahlfors,
   2) all C_5^{-1} non-accretive cubes of $\mathcal{D}$ are disjoint with $F_{\mathcal{D}}$, and
   3) $V_*^\nu(x) \leq C_6$ on $F_{\mathcal{D}}$.

Suppose also that (7.5) holds with constant $\hat{A}$. Let Θ be a core suppression function from (8.36). Then there exists a constant C_4 depending only on $q, C_3, C_5, C_6, \hat{A}$ such that

$$(9.1) \qquad \|V^\Theta : L^2(F_0, \mu) \to L^2(F_0, \mu)\| \leq C_4 \,.$$

The estimate of the norm does not depend on $\kappa > 0$ from (8.36).

Properties of Θ_0.

$$(9.2) \qquad \Theta_0(x) \geq 0\,, \quad \text{and } \Theta \text{ satisfies } (8.4)\,,$$

$$(9.3) \qquad \text{the set } F_0 := \{x : \Theta_0(x) = 0\} \text{ has a "large" measure } \mu(F_0) \geq \beta\,.$$

Indeed, if we consider set E from Lemma 8.9, we see that for every $x \in E$ there exists a set $Z_x \subset \Omega^2$ such that $\theta_\omega(x) = 0$ for all $\omega \in Z_x$ and such that $P\{Z_x\} \geq \beta^2$. Hence $E \subset F_0$. But $\mu(E) \geq \beta\mu(F)$ was proved in Lemma 8.9.

Let C_3 be from Theorem 8.1. Let

$$M := \{y \in \mathbb{R}^n : \text{all balls centered at } y \text{ are } C_3 \text{ Ahlfors balls}\}\,.$$

By the assumption of Theorem 8.1, $F_\omega \subset M$ for every $\omega \in \Omega^2$. Then

$$(9.4) \qquad \forall \omega \in \Omega^2\,, \ \theta_\omega(x) \geq \text{dist}(x, M)\,.$$

Then, by definition,

$$(9.5) \qquad \Theta_0(x) \geq \text{dist}(x, M)\,.$$

Let $R(x)$ denote the supremum of radii of all $2^m C_3$ non-Ahlfors balls centered at x. It is clear that $\text{dist}(x, M) \geq R(x)$. So it is easy to see from (9.4) that

$$(9.6) \qquad \forall \omega \in \Omega^2\,, \ \theta_\omega(x) \geq R(x)\,.$$

By (9.5)

$$(9.7) \qquad \Theta_0(x) \geq R(x)\,.$$

Notice also that

(9.8)   all balls centered at any point $x \in F_0$ are C_3 Ahlfors balls.

In fact, if $x \in F_0$ (that is $\Theta_0(x) = 0$), then for many $\omega \in \Omega^2$, $\theta_\omega(x) = 0$, that is, $x \in F_\omega$ for many ω's; we do not even need many ω's, we need the existence of just one such $\omega(x)$: $x \in F_{\omega(x)}$, because then, by the assumption of Theorem 8.1, all balls centered at x are C_3 Ahlfors balls.

Recall that Theorems 7.1 and 8.1 deal with two measures: μ and $\nu = b\,d\mu$. Let C_6 be from Theorem 8.1. Let $K := \{y \in F : V_*^\nu(x) = V_*(b\,d\mu)(x) \leq C_6\}$. By the assumption of Theorem 8.1, $F_\omega \subset K$ for every $\omega \in \Omega^2$. Then

(9.9) $\forall \omega \in \Omega^2, \ \theta_\omega(x) \geq \operatorname{dist}(x, K)\,.$

Then, by definition,

(9.10) $\Theta_0(x) \geq \operatorname{dist}(x, K)\,.$

In particular, $F_0 = \{x : \Theta_0(x) = 0\} \subset K$, and so

(9.11) $V_*(b\,d\mu)(x) \leq C_6, \ \forall x \in F_0\,.$

Now we can apply Lemma 8.7 to vector-valued kernel $k = (\Psi^1, \dots, \Psi^n)$ (Riesz kernels introduced in Chapter 3). Recall the notation from (8.6) and Section 8.2. Then the operator T with kernel k comes into play, along with its suppressions T^{θ_ω}, T^{Θ_0}, T^Θ. Then (we use (7.5) along with Lemma 8.7 and (9.7), (9.10)):

(9.12) $\forall x \in F, \ T_*^\Theta(b\,d\mu)(x) \leq C_6 + AC_3\hat{A}\,.$

Lemma 8.7 and (9.6), (9.9) give

(9.13) $\forall x \in F, \ T_*^{\theta_\omega}(b\,d\mu)(x) \leq C_6 + AC_3\hat{A}\,.$

(9.14) $\forall x \in F, \ T_*^{\Theta \vee \theta_\omega}(b\,d\mu)(x) \leq C_6 + AC_3\hat{A}\,.$

We are ready to prove Theorem 8.1.

Proof. We are ready to deduce Theorem 8.1 from Theorem 9.1. We need to prove the existence of constants for $C_4 < \infty$, depending on $q, C_3, C_5, C_6, \hat{A}$ (see (7.5)) and $q_0 \in (0, 1)$, and a compact subset F_0 of F such that

(9.15) $\mu(F_0) \geq (1 - q_0)\mu(F)\,,$

(9.16) $\mu(B(x, r)) \leq C_3\, r^{n-1} \ \forall x \in F_0\,,$

(9.17) $\|V^\mu : L^2(F_0, \mu) \to L^2(F_0, \mu)\| \leq C_4\,.$

We just choose F_0 to be the zero set of Θ_0. Then the choice of $\tau = 1 - q$ in Lemma 8.9, and (9.3) prove (9.15) with $q_0 = 1 - \tau/2$.

The kernel $k_{\Theta_0}(x, y)$ of operator T^{Θ_0} is equal to the kernel of operator V, when $x, y \in F_0$. This is just because $\Theta_0 = 0$ on F_0 and by the property of suppressed kernels; for that see (8.8) or Lemma 8.2. But uniformly in κ operator $T^{\Theta_0 + \kappa}$ is bounded from $L^2(F, \mu)$ to itself (by C_4). Let M_{F_0} denote the operator of multiplication of the characteristic function of F_0. Then, uniformly in κ the operators $M_{F_0} T^{\Theta_0 + \kappa} M_{F_0}$ is bounded from $L^2(F_0, \mu)$ to itself. Then $M_{F_0} T^{\Theta_0} M_{F_0}$ is bounded by C_4. From what we said before about the kernel $k_{\Theta_0}(x, y)$ of operator T^{Θ_0}, it follows that

$$M_{F_0} V M_{F_0} = M_{F_0} T^{\Theta_0} M_{F_0}\,.$$

Therefore (9.1) is proved too. As for (9.16), it is (9.8). $\square$

CHAPTER 10

The Fourth Reduction

In this chapter we prove Theorem 9.1, or, better put, we organize one more reduction. This reduction will be the last one. To do it we need the following decomposition.

10.1. μ, b, D, η decomposition

Given a measure μ, a function $b \in L^\infty(\mu)$, and two numbers η, D, we will define a special decomposition of $\varphi \in L^2(\mu)$.

For a constant D, consider

$$M = M(D) := \{x \in F : \text{every ball centered at } x \text{ is } D \text{ Ahlfors}\}.$$

Any such set M is closed.

Let us fix a dyadic lattice $\mathcal{D}$. We are always in the situation when $F = \operatorname{supp}\mu$ and these lattices satisfy (7.19). Let η be finite and positive constant. A cube Q of any of these lattices we call η non-accretive if

$$(10.1) \qquad |\nu(Q)| = |\int_Q b\, d\mu| \le \eta\mu(Q).$$

Notice that cube Q with $\mu(Q) = 0$ is η non-accretive.

Cubes are open. Let $T_\mathcal{D}$ be the unions of all η non-accretive cubes from $\mathcal{D}$. We call $T_\mathcal{D}$ *the non-accretive set* with respect to the lattice. The reader should think that this open set is not too big. Set

$$(10.2) \qquad H := \mathbb{R}^n \setminus M(D).$$

We call the cube Q from $\mathcal{D}$ *a terminal cube*, if $2Q$ is contained in H or Q is contained in $T_\mathcal{D}$ (is η non-accretive).

Let $\mathcal{D}$ be a dyadic lattice above. For a function $\psi \in L^1(\mu)$ and for a cube $Q \subset \mathbb{C}$, denote by $\langle\psi\rangle_Q$ the average value of ψ over Q with respect to the measure μ, i.e.,

$$\langle\psi\rangle_Q := \frac{1}{\mu(Q)}\int_Q \psi\, d\mu$$

(of course, $\langle\psi\rangle_Q$ makes sense only for cubes Q with $\mu(Q) > 0$).

Put

$$\Lambda\varphi := \frac{\langle\varphi\rangle_{Q^0}}{\langle b\rangle_{Q^0}} b.$$

Clearly, $\Lambda\varphi \in L^2(\mu)$ for all $\varphi \in L^2(\mu)$, and $\Lambda^2 = \Lambda$, i.e., Λ is a projection. Note also that actually Λ does not depend on the lattice $\mathcal{D}$ satisfying (7.19), because the average is taken over the whole support of the measure μ regardless of the position of the cube Q^0 (or R^0).

From now on, we will always denote by Q_j ($j = 1, \ldots, 2^n$) the 2^n dyadic subcubes of a cube Q enumerated in some "natural order".

For every transit cube $Q \in \mathcal{D}_1$, define $\Delta_Q \varphi$ by

$$\Delta_Q \varphi\big|_{\mathbb{R}^n \setminus Q} := 0, \qquad \Delta_Q \varphi\big|_{Q_j} := \begin{cases} \left[\dfrac{\langle \varphi \rangle_{Q_j}}{\langle b \rangle_{Q_j}} - \dfrac{\langle \varphi \rangle_Q}{\langle b \rangle_Q} \right] b & \text{if } Q_j \text{ is transit;} \\[2ex] \varphi - \dfrac{\langle \varphi \rangle_Q}{\langle b \rangle_Q} b & \text{if } Q_j \text{ is terminal} \end{cases}$$

($j = 1, \ldots, 2^n$). Observe that for every transit cube Q, we have $\mu(Q) > 0$ and

$$|\langle b \rangle| > \eta$$

for all transit cubes, so our definition makes sense: no zero can appear in the denominator.

10.2. Good functions and bad functions

Suppose now that we have two lattices $\mathcal{D}_1$, $\mathcal{D}_2$ satisfying (7.19). First we define what is a **bad cube** and a **good cube**.

Good and bad cubes. Fix a small number $\delta > 0$. Set

(10.3) r integer $: 2^{-r} \le \delta^S < 2^{-r+1}$,

where S is a large number to be chosen (for now think that it is 2).

By skR we denote $\cup_{i=1}^{2^n} \partial R_i$, where R_i are dyadic children of R.

Let ε, m be parameters from (8.7). Fix forever

(10.4) $\alpha := \dfrac{\varepsilon}{2\varepsilon + 2m}$.

Definition. A cube $Q \in \mathcal{D}_1$ is called *bad* (actually δ-bad) if there exists a cube $R \in \mathcal{D}_2$ such that
  1) $\ell(R) \ge 2^r \ell(Q)$,
  2) $\text{dist}(Q, skR) < \ell(Q)^\alpha \ell(R)^{1-\alpha}$.

Symmetric definition gives us bad R's from $\mathcal{D}_2$. *Good* cubes are those which are not bad.

Notice that the property to be bad does not depend on Q; it depends on how it is located with respect to the second dyadic lattice $\mathcal{D}_2$. The main result is that the probability of being bad is very small.

We consider the probability space Ω^2 of two independent dyadic lattices satisfying (7.19). An element of Ω^2 will be called (ω_1, ω_2) or $(\mathcal{D}_1, \mathcal{D}_2)$.

THEOREM 10.1. *One can choose $S = S(\alpha)$ in such a way that for any fixed $Q \in \mathcal{D}_1$,*

(10.5) $\mathbb{P}_{\omega_2}\{Q \text{ is bad}\} \le \delta^2$.

By symmetry $\mathbb{P}_{\omega_1}\{Q \text{ is bad}\} \le \delta^2$ for any $R \in \mathcal{D}_2$.

Proof. Consider the unit cube Q^0 of $\mathcal{D}$, which satisfies $F \subset \frac{1}{4} Q^0$. It contains F deep inside itself; see (7.19). In particular, $\omega + Q^0$ contains F in $\frac{1}{2}(\omega + Q^0)$ for every $\omega \in (-1/40, 1/40)^n$. The normalized Lebesgue measure on the cube $(-1/40, 1/40)^n$ will be our probability measure. All our lattices "start" with the cube of unit side

$\omega + Q^0$, and then $\omega + Q^0$ is dyadically subdivided, and thus generates the dyadic lattice $\mathcal{D}(\omega)$. We consider two such arbitrary lattices $\mathcal{D}_1$, $\mathcal{D}_2$.

From this point on, we will call a cube $Q \in \mathcal{D}_1$ bad if there exists a cube $R \in \mathcal{D}_2$ such that $\mathrm{dist}(Q, \partial R) \le \ell(Q)^\alpha \ell(R)^{1-\alpha}$ and $\ell(R) \ge 2^m \ell(Q)$.

Choice of S. Fix $k \ge r$. Let us estimate the probability that there exists a cube $R \in \mathcal{D}_2$ of size $\ell(R) = 2^k \ell(Q)$ such that $\mathrm{dist}(Q, \partial R) \le \ell(Q)^\alpha \ell(R)^{1-\alpha}$. It equals t6 the ratio of the area of the narrow strip around the boundary of the cube R of size $2^k \ell(Q)$ to the whole area of $2^k Q$. The width of the strip is bounded by $\ell(Q) + 2^{(1-\alpha)k} \ell(Q)$. We conclude that this ratio is less than $\widetilde{A} \left[\frac{\ell(Q)}{\ell(R)} \right]^\alpha = \widetilde{A} \cdot 2^{-k\alpha}$.

Therefore the probability that the cube Q is bad does not exceed

$$\widetilde{A} \sum_{k=r}^\infty 2^{-k\alpha} = \frac{\widetilde{A} \cdot 2^{-r\alpha}}{1 - 2^{-\alpha}} \,.$$

Now the reader is asked to take a look at (10.3).

We choose a minimal S such that $\widetilde{A} \frac{\delta^{S\alpha}}{1 - 2^{-\alpha}} \le \delta^2$ (of course, $S = 3/\alpha$ is enough for all small δ). $\qquad\square$

Let us consider another possible definition of bad cubes.

Definition. A cube $Q \in \mathcal{D}_1$ is called *bad in the second sense* (actually δ-bad in the second sense) if there exists a cube $R \in \mathcal{D}_2$ such that 1) $\ell(R) \ge 2^r \ell(Q)$, 2) $\mathrm{dist}(Q, skR) < \ell(Q)^\alpha \ell(R)^{1-\alpha}$, or if it is *badly intersected*. A cube $Q \in \mathcal{D}_1$ is called *badly intersected* if there exists a cube $R \in \mathcal{D}_2$ such that 1) $2^{-r} \ell(Q) \le \ell(R) \le 2^r \ell(Q), R \subset 5 \cdot 2^r Q$, 2) there exists a boundary hyperplane P of R such that

$$(10.6) \qquad \exists t : \mu(\{x \in Q : \mathrm{dist}(x, P) \le t\}) > \frac{\mu(Q)}{2^{-2r}\delta} \frac{t}{\mathrm{diam}(Q)} \,.$$

Symmetric definition gives us bad R's from $\mathcal{D}_2$. *Good in the second sense* cubes are those which are not bad in the second sense. We skip the words "in the second sense" if it does not matter for us which definition to use. If we need to emphasize that, say, only the second definition is applicable, we specifically mention this.

THEOREM 10.2. *One can choose $S = S(\alpha)$ in such a way that for any fixed $Q \in \mathcal{D}_1$*

$$(10.7) \qquad \mathbb{P}_{\omega_2}\{Q \text{ is bad in the second sense}\} \le 2\delta^2 \,.$$

By symmetry $\mathbb{P}_{\omega_1}\{Q \text{ is bad in the second sense}\} \le \delta^2$ *for any* $R \in \mathcal{D}_2$.

The proof is the same as for Theorem 10.1. We will sketch it. Notice that the positions of hyperplanes satisfying (10.6) are very few. Let us explain this. Fix a normal direction to one of the faces of cubes of our dyadic lattice (coordinate ort, say e_1). Consider all hyperplanes with this normal. Consider those that intersect Q; call this family H. Notice that (10.6) is scale invariant. So one can consider Q of the unit size. Project $\mu|Q$ onto the direction 6f the line L_1 of direction e_1, that is along hyperplanes of family H. Projected measure is, say, μ_1 on L_1. If hyperplane $h \in H$ satisfies (10.6), then the point $x \in L_1$ satisfies

$$\exists t : \mu_1(\{y \in L_1 : |y - x| \le t\}) > \frac{\mu(Q)}{2^{-2r}\delta} t \,.$$

By the weak type theorem for the maximal function

$$M\mu_1(x) := \sup \frac{\mu_1(\{y \in L_1 : |y - x| \le t\})}{t} \,,$$

the set of x's can only have measure bounded by $A\,\delta \cdot 2^{-2r}$. We can repeat this for any other coordinate vector. There are n of them. So finally let us call the set of positioning of hyperplanes intersecting Q (parallel to one of the faces of Q) and satisfying (10.6) PH. There is a natural Lebesgue measure on them and we have just proved that

$$|HP|/\ell(Q) \le A\,\delta \cdot 2^{-2r}\,.$$

Now it is clear that for a cube R from $\mathcal{D}_2$ and of size coinciding with the size of Q, the probability to have one of its faces in HP is at most $A\,\delta \cdot 2^{-2r}$. But we should also consider R's of the type $\ell(R) = 2\,\ell(Q), 2^2\,\ell(Q), \ldots, 2^r\ell(Q)$ as well as $\ell(R) = 2^{-1}\,\ell(Q), 2^{-2}\,\ell(Q), \ldots, 2^{-r}\ell(Q)$. For the "larger" cubes the probability to have one of its faces in HP is at most $A\,\delta \cdot 2^{-2r}$ again. For a cube R such that $\ell(R) = 2^{-r}\ell(Q)$ the probability to have one of its faces in HP is at most $A\,\delta \cdot 2^{-r}$. The same holds true for larger cubes. Summing all up we get that the probability to be badly intersected is at most $2r\,2^{-r}\,\delta$. This of course is not greater than δ^2 as soon as $S > 10$, and δ is sufficiently small (see (10.3)).

Now let us take a look at μ, b, D, η decomposition from Section 10.1. We fix two lattices $\omega = (\mathcal{D}_1, \mathcal{D}_2) \in \Omega^2$. Apply to $f \in L^2(\mu)$ by μ, b, D, η decomposition with respect to $\mathcal{D}_1$.

Definition. Function f is called *good* if $\Delta_Q(f) = 0$ for all bad $Q \in \mathcal{D}_1$. Function f is called *bad* if $\Delta_Q(f) = 0$ for all good $Q \in \mathcal{D}_1$. Similarly we define good and bad functions if decomposition is made with respect to the second dyadic lattice of ω. There will be no misunderstanding of which decomposition we mean.

Always

$$f = f_{good} + f_{bad}\,.$$

THEOREM 10.3. *Let function b satisfy* (7.5). *We consider* μ, b, D, η *decomposition of* f, *and take a bad part of it for every* $\omega = (\omega_1, \omega_2) \in \Omega^2$. *One can choose* $S = S(\alpha)$ *in such a way that*

(10.8) $$\mathbb{E}(\|f_{bad}\|) \le C(\hat{A}, \eta)\delta\|f\|\,.$$

We postpone the proof of this theorem untill Section 13.1.

10.3. Estimates of nonhomogeneous Calderón–Zygmund operators on good functions

We are given a positive measure μ with compact support $F \subset \mathbb{R}^n$. Let k be a vector valued function satisfying (8.7). We always consider only *bounded kernels* k. Then k defines a bounded operator on $L^2(\mu)$, denoted by T. Number m from (8.7) is called the order of k or T. Actually, we need only $m = n - 1$, but this does not matter for the results of this section.

We will say that k has Calderón–Zygmund constant C if constants from (8.7) are bounded by C.

Let first μ be also *of order m*, namely, let

(10.9) $$\mu(B(x,r)) \le B\,r^m, \ \ \forall B(x,r)\,.$$

In other words all balls are B Ahlfors balls. We write down the succession of the theorems; each is more general than the previous one. We will need to use the last and most difficult one. But the previous ones serve here the purpose of leading us smoothly to this last one.

Notation.

$$\langle \varphi, \psi \rangle := \int_F \varphi \psi \, d\mu \, .$$

Operator T_* was defined in (8.29).

THEOREM 10.4. *Let k be a continuous Calderón–Zygmund kernel with constant C and of order $m \leq n$. Let*

(10.10) $$T_* \mathbf{1} \leq C_6 \, .$$

Let all balls be B Ahlfors balls- in other words, we have (10.9). Let $\mathcal{D}_i$, $i = 1, 2$ be two dyadic lattices and f_{good}, g_{good} are two δ-good in the second sense functions with respect to $\mathcal{D}_1$ and $\mathcal{D}_2$, respectively. Then there exists C_7 depending only on B, C, m, n, C_6 such that

(10.11) $$|\langle T f_{good}, g_{good} \rangle| \leq A \, C_7 \, \delta^{-A} \|f_{good}\| \|g_{good}\| \, .$$

The next theorem is more general. It answers the question "what we can do if (10.9) does not hold?" In other words, what can be done if not all balls are B Ahlfors balls? We denote, as we have already done many times above, by $R(x)$ denotes the supremum of radii of all B non-Ahlfors balls centered at x. We have the following result.

THEOREM 10.5. *Let k be a continuous Calderón–Zygmund kernel with constant C and of order $m \leq n$. Let*

(10.12) $$T_* \mathbf{1} \leq C_6 \, .$$

Let k satisfy

(10.13) $$|k(x, y)| \leq \min \left[\frac{1}{R(x)^m}, \frac{1}{R(y)^m} \right],$$

where $R(x)$ denotes the supremum of radii of all B non-Ahlfors balls centered at x. Let $\mathcal{D}_i$, $i = 1, 2$ be two dyadic lattices and f_{good}, g_{good} are two δ-good in the second sense functions with respect to $\mathcal{D}_1$ and $\mathcal{D}_2$, respectively. Then there exists C_7 depending only on B, C, m, n, C_6 such that

(10.14) $$|\langle T f_{good}, g_{good} \rangle| \leq A \, C_7 \, \delta^{-A} \|f_{good}\| \|g_{good}\| \, .$$

Function $\mathbf{1}$ is of course an accretive one. The next theorem is again more general. It answers the question "what we can do if (10.9) does not hold and if instead of an accretive function we consider a non-accretive one?" Let us fix two dyadic lattices $\mathcal{D}_1$ and $\mathcal{D}_2$. That means that we fixed an $\omega \in \Omega^2$ introduced in Section 8.4. We are always in the situation when $E = \operatorname{supp} \mu$ and these lattices satisfy (7.19). Let C_5 be finite and positive constant. A cube Q of any of these lattices we call C_5^{-1} non-accretive if (7.21) is false. Cubes are open. Set $T_{\mathcal{D}_1}, T_{\mathcal{D}_2}$ to be unions of all C_5^{-1} non-accretive cubes from $\mathcal{D}_1$ and $\mathcal{D}_2$ correspondingly. Let $T_{12} := T_{\mathcal{D}_1} \cup T_{\mathcal{D}_1}$. We call T_{12} *the non-accretive set* with respect to this pair of lattices. The reader should think that this open set is not too large.

THEOREM 10.6. *Let $k = k_\omega$ be a continuous Calderón–Zygmund kernel with constant C and of order $m \leq n$. Let b satisfy (7.5) with constant $\hat{A}$, and for operator $T = T_\omega$ with kernel k_ω, the following holds:*

(10.15) $$T_* b \leq C_6 \, .$$

Let k satisfy

(10.16)
$$|k(x,y)| \leq \min\left[\frac{1}{R(x)^m}, \frac{1}{R(y)^m}\right],$$

where $R(x)$ denotes the supremum of radii of all B non-Ahlfors balls centered at x. Let k satisfy

(10.17)
$$|k(x,y)| \leq \min\left[\frac{1}{d(x)^m}, \frac{1}{d(y)^m}\right],$$

where $d(x) := \operatorname{dist}(x, \mathbb{R}^n \backslash T_{12})$. There exists C_8 depending only on B, C, m, n, C_5, C_6 (and not depending on T, f, g) such that for every positive δ,

(10.18)
$$|\langle Tf, g \rangle| \leq A\, C_8\, \delta^{-A} \|f\| \|g\| + \|T\| R(\omega, f, g),$$

where $R(\omega, f, g)$ does not depend on T and

(10.19)
$$\mathbb{E}R(\omega, f, g) \leq A\delta \|f\| \|g\|.$$

This theorem will now play the leading part in proving our latest reduction, Theorem 9.1.

Remark on terminology. The first two theorems of this section are nonhomogeneous $T1$ theorems; the last one is a *nonhomogeneous non-accretive Tb* theorem. The word *nonhomogeneous* is used here because we deal with a nonhomogeneous measure μ. The L^2 theory and corresponding Tb theorems for Calderón–Zygmund operators on spaces of homogeneous type were proved by Christ in [**5**]. The L^p theory of Calderón–Zygmund operators on homogeneous spaces was established by Coifman and Weiss. They noticed that classical L^p theory of Calderón–Zygmund operators with respect to Lebesgue measure can be generalized to the case of doubling measure. Metric measure spaces with doubling measure are often called homogeneous spaces.

10.4. The reduction of Theorem 9.1 to estimates of nonhomogeneous Calderón–Zygmund operator, namely to Theorem 10.6

We will need operators
$$T^{\Theta}, T^{\theta_\omega}, T^{\Theta \vee \theta_\omega}, C^{\Theta}, C^{\theta_\omega}, C^{\Theta \vee \theta_\omega}$$
and their kernels
$$k_{\Theta}, k_{\theta_\omega}, k_{\Theta \vee \theta_\omega}, c_{\Theta}, c_{\theta_\omega}, c_{\Theta \vee \theta_\omega}$$
introduced in Chapter 8. The definitions of these kernels and operators can be found in Section 8.1. The base kernel k used in these definitions can be any antisymmetric Calderón–Zygmund kernel of order m. We will apply everything only to $k = (\Psi^1, \ldots, \Psi^n), \Psi^i := \frac{x_i - y_i}{|x-y|^n}$, that is, to the Riesz kernel of order $m = n - 1$ introduced in Section 3.

Notation. Let positive measure μ be given. We introduce also the following maximal functions:
$$M_{m,R}f(x) := \sup_{r > R} \frac{1}{r^m} \int_{B(x,r)} |f(y)| \, d\mu(y).$$
$$M_\mu f(x) := \sup_{r > 0} \frac{1}{\mu(B(x,3r))} \int_{B(x,r)} |f(y)| \, d\mu(y)$$

We will need several lemmas; the first of them is Cotlar's inequality for nonhomogeneous Calderón–Zygmund operators.

We denote, as we have already done many times above, by $R(x)$ the supremum of radii of all B non-Ahlfors balls centered at x. We have the following result.

LEMMA 10.7. *Let k be a continuous Calderón–Zygmund kernel with constant C and of order $m \le n$, and let μ be a positive measure. Let k satisfy*

$$(10.20) \qquad |k(x,y)| \le \min[\frac{1}{R(x)^m}, \frac{1}{R(y)^m}],$$

where $R(x)$ is the supremum of radii of all B non-Ahlfors balls centered at x.

As usual T_* is defined as in (8.29). Fix $p \in (1,2)$. Let $\|T\|$ denote the norm of the operator in $L^2(\mu)$. Then for every x,

$$(10.21) \qquad \begin{aligned} (T_* f)(x) &\le A\,C\,(M_\mu(Tf))(x) + A\,B\,C(M_\mu|f|^p)^{1/p}(x) \\ &\quad + A\,\|T\|\,(M_\mu|f|^p)^{1/p}(x). \end{aligned}$$

We will prove this lemma in the next section.
We need two more kernels.

Notation. Let

$$(10.22) \qquad c(x,y) := \mathbf{E}c_{\Theta \vee \theta_\omega}(x,y), \ \ K := \mathbf{E}k_{\Theta \vee \theta_\omega}(x,y),$$

where $\mathbf{E}$ is the mathematical expectation in our space Ω^2 of two independent lattices satisfying (7.19) and introduced in Section 8.3.

In defining function Θ a certain positive number β was involved; see (8.35).

LEMMA 10.8. *Let Θ be defined as in (8.36). Then for any f,*

$$(10.23) \qquad |C^\Theta f(x)| \le A\,\beta^{-2} c_* f(x) + A\,M_\mu f(x),$$

$$(10.24) \qquad |C^{\Theta \vee \theta_\omega} f(x)| \le A\,\beta^{-2} c_* f(x) + A\,M_\mu f(x).$$

Proof. Let $\mathbf{P}$ denote the probability on Ω^2. Set

$$v_x(t) := \mathbf{P}\{\omega \in \Omega^2 : (\Theta \vee \theta_\omega)(x) \le t\}.$$

It is obvious that
1) $t < \Theta(x) \Rightarrow v_x(t) = 0$.
2) $v_x(t)$ is increasing in t.
3) $t \ge \Theta(x) \Rightarrow v_x(t) \ge \beta^2$.
The last implication follows from the fact that for every x there is a set $Z_x \subset \Omega^2$ such that $\mathbf{P}\{Z_x\} \ge \beta^2$ and $(\Theta \vee \theta_\omega)(x) \le \Theta(x) \le t$ for every $\omega \in Z_x$. Obviously $v_x(t) \ge \mathbf{P}\{Z_x\}$.

Let us look at (10.23). By the definition of c in (10.22) we see

$$c(x,y) = k(x,y)v_x(|x-y|) = v_x(|x-y|)\chi_{\mathbb{R}^n \setminus B(x,\Theta(x))}k(x,y) = v_x(|x-y|)c_\Theta(x,y),$$

which can be written as

$$(10.25) \qquad c_\Theta(x,y) = \beta^{-2}\chi_{\mathbb{R}^n \setminus B(x,\Theta(x))}c(x,y)\varphi(|x-y|),$$

where $\varphi(t) = \frac{\beta^2}{v_x(t)} \le 1$ if $t \ge \Theta(x)$ and is equal to 1 otherwise. Notice that φ is decreasing. We need now the following

LEMMA 10.9. *Let $b(x,y)$ be a kernel such that $|b(x,y)| \leq \frac{1}{|x-y|^m}$. Then we have a well-defined $(b^*f)(x) := \sup_{r>0} | \int_{|y-x|>r} b(x,y)f(y)\,d\mu(y)|$. Let $R > 0$ and let ϕ be a decreasing function on $[0,\infty)$, $0 \leq \phi \leq 1$. Consider*

$$(b_R^\phi f)(x) := \Big| \int_{|y-x|>R} b(x,y)\phi(|x-y|)f(y)d\mu(y) \Big| \,.$$

Then

$$(b_R^\phi f)(x) \leq 2\,(b^*f)(x) + 2\,(M_{m,R}f)(x)\,.$$

Proof. Consider annuli $A_k(x) = \{y : 2^{k-1}R \leq |y-x| \leq 2^k R\}$. Then

$$(b_R^\phi f)(x) \approx \Sigma_{k \geq 1} \int_{A_k} b(x,y)\phi_k f(y)d\mu(y)\,,$$

where ϕ_k are some values (say, left end point values) of $\phi(t)$ for $t \in [2^{k-1}R, 2^k R]$, $k = 1, 2, \ldots$. More precisely ($\phi_0 := 0$),

$$(b_R^\phi f)(x) = \Sigma_{k \geq 1}(\phi_k - \phi_{k-1}) \int_{|y-x| \geq 2^{k-1}R} b(x,y)f(y)\,d\mu(y) + \text{Discrepancy}\,.$$

Thus, the monotonicity of ϕ implies

$$|\text{The first term}| \leq \phi_1 \Big| \int_{|y-x| \geq R} b(x,y)f(y)\,d\mu(y) \Big|$$

$$+ \Sigma_{k \geq 2}(\phi_{k-1} - \phi_k)| \int_{|y-x| \geq 2^{k-1}R} b(x,y)f(y)\,d\mu(y)| \leq 2\,(b^*f)(x)\sup\phi\,.$$

On the other hand, let us use the symbol J_k to denote the jump (the oscillation) of the monotone function ϕ on the interval $[a_k, a_{k+1}]$. Then

$$|\text{Discrepancy}| \leq \Sigma_{k \geq 1} J_k \frac{1}{(2^{k-1}R)^m} \int_{B(x,2^k R)} |f(y)|\,d\mu(y)\,.$$

We continue the previous estimate as follows:

$$|\text{Discrepancy}| \leq 2\,(M_{m,R}f)(x)\Sigma_{k \geq 1} J_k\,.$$

But ϕ was assumed to be monotone and $0 \leq \phi \leq 1$, so the sum of the jumps is bounded by 1. Lemma 10.9 is proved. □

We can continue the proof of Lemma 10.8. Use (10.25) and Lemma 10.9 to conclude (10.23).

But from (10.25) it follows trivially that

$$(10.26) \qquad c_{\Theta \vee \theta_\omega}(x,y) = \beta^{-2}\chi_{\mathbb{R}^n \setminus B(x,\Theta \vee \theta_\omega(x))}c(x,y)\varphi(|x-y|)\,,$$

with the same function φ. Now we apply Lemma 10.9 to this equality. We get then (10.24). Lemma 10.8 is entirely proved. □

To continue the reduction of Theorem 9.1 let us use again the notation $R(x)$ for the supremum of radii of $2C_3$ non-Ahlfors balls. Let us recall Lemma 8.3 and (9.6), (9.7).

In (10.22) we introduced two bounded kernels c and K. Corresponding integral operators (with respect to measure μ) will be called $\mathcal{C}$, $\mathcal{T}$. If we apply Lemma 8.3 (where θ will be played by $\Theta \vee \theta_\omega$, and where $(\Theta \vee \theta_\omega)(x) \geq R(x)$ because of (9.6)), we get

$$(10.27) \qquad |(T^{\Theta \vee \theta_\omega} - C^{\Theta \vee \theta_\omega})f(x)| \leq A\,C_3\,M_\mu f(x)\,.$$

We can take the mathematical expectation $\mathbf{E}$ of (10.27) and get

$$(10.28) \qquad |(\mathcal{T} - \mathcal{C})f(x)| \le A\,C_3\,M_\mu f(x)\,.$$

Using the remark after Lemma 8.3 we can also write for arbitrary positive ε

$$(10.29) \qquad |(T_\varepsilon^{\Theta\vee\theta_\omega} - C_\varepsilon^{\Theta\vee\theta_\omega})f(x)| \le A\,C_3\,M_\mu f(x)\,.$$

Averaging over Ω^2 gives

$$(10.30) \qquad |(\mathcal{T}_\varepsilon - \mathcal{C}_\varepsilon)f(x)| \le A\,C_3\,M_\mu f(x)\,.$$

By the definition of star operation, inequality (10.30) implies

$$(10.31) \qquad |(\mathcal{T}_* - \mathcal{C}_*)f(x)| \le A\,C_3\,M_\mu f(x)\,.$$

We are in a position to prove Theorem 9.1. Having in mind that all reductions were already made, our Theorem 8.1 and 7.1 will also be proved. But in the proof below we will use Lemma 10.7 and Theorem 10.6. These results will be proved in the next sections. So, now we prove Theorem 9.1.

Proof. We fix a point $\omega \in \Omega^2$. As $\omega = (\omega_1, \omega_2)$, and each ω_i stands for a dyadic lattice having property (7.19), we automatically have fixed two such dyadic lattices $\mathcal{D}_1, \mathcal{D}_2$.

We fix also two functions $f, g \in L^2(\mu)$. We will use operator M_μ. It is known that it is bounded on $L^2(\mu)$, and its norm in this space will be denoted by $\|M_\mu\|$.

By (10.27)

$$|\langle C^{\Theta\vee\theta_\omega}f, g\rangle| \le |\langle T^{\Theta\vee\theta_\omega}f, g\rangle| + AC_3\,\|M_\mu\|\|f\|\|g\|\,.$$

We see that we now have to estimate $|\langle T^{\Theta\vee\theta_\omega}f, g\rangle|$ using Theorem 10.6:

$$|\langle T^{\Theta\vee\theta_\omega}f, g\rangle| \le AC_8\delta^{-A}\|f\|\|g\| + \langle T^{\Theta\vee\theta_\omega}\|R(\omega, f, g)\,.$$

Now we estimate $\langle T^{\Theta\vee\theta_\omega}\|$ by (10.27) followed by Lemma 10.8 (namely by (10.24)):

$$\langle T^{\Theta\vee\theta_\omega}\| \le \beta^{-2}\|C_*\| + A\|M_\mu\|\,.$$

Combine the last two inequalities into

$$|\langle T^{\Theta\vee\theta_\omega}f, g\rangle| \le AC_8\delta^{-A}\|f\|\|g\| + (\beta^{-2}\|C_*\| + A\|M_\mu\|)R(\omega, f, g)\,.$$

Recall that $\mathcal{T} = \mathbf{E}T^{\Theta\vee\theta_\omega}$ (and $\mathcal{C} = \mathbf{E}C^{\Theta\vee\theta_\omega}$). Average the last inequality and use then (10.31) to get

$$(10.32) \quad |\langle \mathcal{T}f, g\rangle| \le A\,C_8\,\delta^{-A}\|f\|\|g\| + A\,\beta^{-2}\,\delta\|\mathcal{T}_*\|\|f\|\|g\| + A'\,\|M_\mu\|\|f\|\|g\|$$

or

$$(10.33) \qquad \|\mathcal{T}\| \le A\,C_8\,\delta^{-A} + A\,\beta^{-2}\,\delta\|\mathcal{T}_*\| + A'\,\|M_\mu\|\,.$$

To continue the proof of Theorem 9.1 let us use its notation; namely, let us use again the notation $R(x)$ for the supremum of radii of $2C_3$ non-Ahlfors balls. Let us recall (9.6). In particular, the kernels $k_{\Theta\vee\theta_\omega}(x, y)$ are majorized by $\min[1/R(x)^m, 1/R(y)^m]$. Therefore, the same holds for the averaged kernel K from (10.22). We just verified the assumption of Lemma (10.7) for Calderón–Zygmund operator $\mathcal{T}$ with bounded kernel K. If we apply Lemma 10.7 then (10.33) implies (C_9 depends on C_3 and on Calderón–Zygmund parameters of $\mathcal{T}$)

$$(10.34)$$
$$\|\mathcal{T}_*\| \le C_9(\|\mathcal{T}\| + \|M_\mu\|) \le A'\,C_9\,C_8\,\delta^{-A} + A''\,\beta^{-2}\,C_9\delta\|\mathcal{T}_*\| + A'''\,C_9\,\|M_\mu\|\,.$$

Choosing

(10.35) $\delta = a\beta^2$, where a is sufficiently small depending on C_3, C_9 ,

we get from (10.34) that

(10.36) $\|\mathcal{T}_*\| \leq C(C_3, C_8, n)\beta^{-A}\|M_\mu\|$.

In conjunction with (10.31), this gives (with different A's)

(10.37) $\|\mathcal{C}_*\| \leq C(C_3, C_9, n)\beta^{-A}\|M_\mu\|$.

Finally we use Lemma 10.8 once more. Namely, we apply (10.23). Then (10.23) and (10.37) together give

(10.38) $\|C^\Theta\| \leq A'\, C(C_3, C_9, n)\beta^{-A-2}\|M_\mu\|$.

We are left to use (9.7): $\Theta(x) \geq R(x)$ in conjunction with Lemma 8.3, to conclude

(10.39) $\|K^\Theta\| \leq A'\, C(C_3, C_9, n)\beta^{-A-2}\|M_\mu\| + A''\, C_3\|M_\mu\|$.

Notice that the number β from the definition (8.35) of function Θ was introduced as (see (8.34) in Section 8.4)

$$\beta = \frac{1-q}{2} ,$$

where q is the parameter from Theorem 9.1. Setting

$$C_4 = A'\, C(C_3, C_9, n)\beta^{-A-2}\|M_\mu\| + A''\, C_3\|M_\mu\|,$$

we get exactly (9.1), which means that Theorem 9.1 is proved. $\square$

    To make the proof self-contained we need to prove Theorem 10.6 and Lemma 10.7, which was used before 10.34. The second task is easier, and we start with it.

CHAPTER 11

The Proof of Nonhomogeneous Cotlar's Lemma. Arbitrary Measure

In Section 5.3 we already saw nonhomogeneous Cotlar's lemma. But in Section 5.3, underlying measure had a restriction (10.9); namely, it was of order m. But we need Cotlar's inequality for measures without this assumption.

For every $x \in \operatorname{supp} \mu$, let $\mathcal{R}(x)$ denote the supremum of R such that $B(x, R)$ is M non-Ahlfors ball.

Let the kernel k of T be an asymmetric Calderón–Zygmund kernel of order m (see (8.7)) satisfying

$$(11.1) \qquad |k(x, y)| \leq \min \left[\frac{1}{\mathcal{R}(x)^m}, \frac{1}{\mathcal{R}(y)^m} \right].$$

Let us recall the notation. Given positive measure μ with compact support, symbol $T^* = T_\mu^*$ denotes maximal singular operator. Let ν be an arbitrary positive measure. The maximal operator natural for non-doubling measure is

$$M_\mu \nu(x) := \sup_{r > 0} \frac{\nu(B(x, r))}{\mu(B(x, 3r))} .$$

Let k satisfy (11.1). Suppose T with kernel k is bounded in $L^2(\mu)$. Let $\beta > 0$; f we define also

$$M_{\mu, \beta} f(x) := \sup_{r > 0} \left(\frac{1}{\mu(B(x, 3r))} \int_{B(x, r)} |f(y)|^\beta \, d\mu(y) \right)^{1/\beta} .$$

Then we want to prove the following theorem.

THEOREM 11.1. *Let μ be an arbitrary positive measure. Let $\mathcal{R}(x)$ denote the supremum of R such that $B(x, R)$ is M non-Ahlfors ball. Let k satisfy (11.1). Suppose T with kernel k is bounded in $L^2(\mu)$. Let $\beta \in (1, 2)$. Then*

$$(11.2) \qquad T^* f(x) \leq C_1 \, M_\mu T f(x) + C_2 \, M_{\mu, \beta} f(x) .$$

Constant C_1 depends on n, m, and Calderón–Zygmund constants of the kernel; constant C_2 depends in addition on $\|T\|_2 := \|T\|_{L^2(\mu) \to L^2(\mu)}$ and on constant M, which defined $\mathcal{R}(x)$ from (11.1).

Proof. The proof follows the proof of Theorem 7.1 on pages 479-480 of [**44**] in line after line fashion. Recall that the proof in [**44**] consists of the following steps:

    1) a lemma of Guy David,

    2) weak type estimate for T,

    3) $L^\beta(\mu)$ estimate for T (just by interpolation between $L^2(\mu)$ and weak L^1 using $\beta \in (1, 2)$).

We go through all of these steps here again. Notice the difference in notation: T^* was denoted $T^\sharp$ in [**44**]. And T^* in [**44**] is the operator with transposed kernel $k(y,x)$.

We start the proof by fixing $r > 0, x \in \operatorname{supp}\mu$, and putting $\hat{r}^m = \max[r, \mathcal{R}(x)]$. Consider

$$(T^r f)(x) := \int_{y:|y-x|\geq r} k(x,y) f(y) \, d\mu(y) \,.$$

Put

$$r_j^m := 3^j \hat{r}^m, \mu_j := \mu(B(x, r_j^m)) \,.$$

Let k be the smallest index such that $\mu_{k+1} \leq 36\mu_{k-1}$. It exists, because otherwise, for every k, $\mu(B(x, \hat{r}^m)) \leq 36^{-k}\mu_{2k} \leq 2M36^{-k}r_{2k}^m$. This is because our radii are greater than $\mathcal{R}(x) := \sup\{r > 0 : \mu(B(x,r)) > Mr^m\}$. We continue with $\mu(B(x, \hat{r}^m)) \leq 2M36^{-k}3^{2k}\hat{r}^m = 2M2^{-2k}\hat{r}^m$. This contradicts the assumption $x \in \operatorname{supp}\mu$.

Let $R := r_{k-1}^m$. We estimate

$$|(T^r f)(x) - (T^{3R} f)(x)| \leq \int_{B(x,\hat{r}^m)\setminus B(x,r)} |k(x,y)| \, |f(y)| \, d\mu(y)$$

$$+ \sum_{j=1}^{k} \int_{B(x,r_j^m)\setminus B(x,r_{j-1}^m)} \cdots \,.$$

The first term vanishes if $\hat{r}^m > \mathcal{R}(x)$. Otherwise it is bounded by

$$\frac{1}{\mathcal{R}(x)^m} \int_{B(x,\hat{r}^m)} |f(y)| \, d\mu(y) = \frac{1}{\hat{r}^{m^2}} \int_{B(x,\hat{r}^m)} |f(y)| \, d\mu(y)$$

$$\leq \frac{\mu(B(x, 3\hat{r}^m))}{\hat{r}^{m^2} \mu(B(x, 3\hat{r}^m))} \int_{B(x,\hat{r}^m)} |f(y)| \, d\mu(y) \,.$$

And this is less than $AM\, M_\mu f(x)$. Similarly,

$$\int_{B(x,r_j^m)\setminus B(x,r_{j-1}^m)} |k(x,y)| \, |f(y)| \, d\mu(y) \leq \frac{\mu_{j+1}}{r_{j-1}^m \mu(B(x, r_{j+1}^m))} \int_{B(x,r_j^m)} |f(y)| \, d\mu(y) \,.$$

But we know that $r_{j-1}^m = 3^{-k+j-1}r_k^m, \mu_{j+1} \leq 36(36)^{\frac{-k+j}{2}}\mu_k$. Hence $\frac{\mu_{j+1}}{r_{j-1}^m}$ is bounded by $36 \cdot 3^{k-j+1}6^{-k+j}\frac{\mu_k}{r_k^m} \leq AM2^{-k+j}$. Therefore,

$$\sum_{j=1}^{k} \int_{B(x,r_j^m)\setminus B(x,r_{j-1}^m)} |k(x,y)| \, |f(y)| \, d\mu(y)$$

$$\leq AM \sum_{j=1}^{k} 2^{-k+j} \frac{1}{\mu(B(x, r_{j+1}^m))} \int_{B(x,r_j^m)} |f(y)| \, d\mu(y) \,.$$

The last sum is obviously bounded by $AM\, M_\mu f(x)$. We finally get

$$|(T^r f)(x) - (T^{3R} f)(x)| \leq AM\, M_\mu f(x) \,.$$

Now we need to estimate $(T^{3R} f)(x)$. Consider the average

$$V_R(x) := \frac{1}{\mu(B(x,R))} \int_{B(x,R)} Tf \, d\mu \,.$$

First,

$$|V_R(x)| \leq \frac{\mu(B(x,3R))}{\mu(B(x,R))} M_\mu[Tf](x) \leq 36 M_\mu[Tf](x) \,.$$

Second,

$$V_R(x) - (T^{3R}f)(x) = \int_{\mathbb{R}^n \setminus B(x,3R)} T'[\delta_x - \frac{1}{\mu(B(x,R))}\chi_{B(x,R)}\, d\mu]f\, d\mu$$

$$- \frac{1}{\mu(B(x,R))}\int_{B(x,R)} T[f\chi_{B(x,3R)}]\, d\mu = I + II \,.$$

Here T' denotes the operator with kernel $k(y,x)$.

Estimate of I. Put $\eta = \delta_x - \frac{1}{\mu(B(x,R))}\chi_{B(x,R)}\, d\mu$. All radii greater than $3R$ are M-Ahlfors for μ. This and the fact that $\eta(\mathbb{C}) = 0$ allows us to use Lemma 11.4. So $I \leq A\,M\,\|\eta\|\,M_\mu f(x) \leq A\,M\,M_\mu f(x) \leq A\,M\,M_{\mu,\beta}f(x)$.

Estimate of II. Fix $\beta \in (1,2)$. Let $1/\alpha + 1/\beta = 1$.

$$|II| \leq \frac{1}{\mu(B(x,R))}\|\chi_{B(x,R)}\|_{L^\alpha(\mu)}\|T(f\chi_{B(x,3R)})\|_{L^\beta(\mu)} \leq \|T\|_\beta \frac{(\int_{B(x,3R)}|f|^\beta\, d\mu)^{\frac{1}{\beta}}}{\mu(B(x,R))^{\frac{1}{\beta}}} \,.$$

Here we abbreviate $\|T\|_\beta := \|T\|_{L^\beta(\mu)\to L^\beta(\mu)}$. We can continue

$$|II| \leq \|T\|_\beta \frac{\mu(B(x,9R))^{\frac{1}{\beta}}(M_{\mu,\beta}f)(x)}{\mu(B(x,R))^{\frac{1}{\beta}}}$$

$$\leq 36^{\frac{1}{\beta}}\|T\|_\beta(M_{\mu,\beta}f)(x) \leq A\|T\|_\beta(M_{\mu,\beta}f)(x) \,.$$

To estimate $\|T\|_\beta$ via $\|T\|_2$ we first need the following estimate.

Estimate of weak type via $\|T\|_2$. Here is David's lemma. The only difference is that measure is not assumed to be of order m, but the proof is the same.

LEMMA 11.2. *Let the kernel k of T satisfy (11.1). For any measurable set F and any point $x \in \operatorname{supp}\mu$,*

$$T^*\chi_F(x) \leq A_1 M_\mu[T\chi_F](x) + A_2 M + A_3\|T\|_2 \,.$$

Proof. Fix $x \in \operatorname{supp}\mu$ and $r > 0$. Put $\hat{r}^m = \max[r, \mathcal{R}(x)]$, where $\mathcal{R}(x) := \sup\{r > 0 : \mu(B(x,r)) > Mr^m\}$. Consider $r_j^m = 3^j\hat{r}^m$. Put $\mu_j := \mu(B(x,r_j^m))$. Let k be the smallest index such that $\mu_k \leq 6\mu_{k-1}$. It exists. Otherwise, for every k, $\mu(B(x,\hat{r}^m)) \leq 6^{-k}\mu_k \leq 2M6^{-k}r_k^m$. This is because our radii are greater than $\mathcal{R}(x)$. We continue with $\mu(B(x,\hat{r}^m)) \leq 2M6^{-k}3^k\hat{r}^m = 2M2^{-k}\hat{r}^m$. This contradicts the assumption $x \in \operatorname{supp}\mu$. Put $R = r_{k-1}^m$. We estimate

$$|(T^rf)(x) - (T^{3R}f)(x)| \leq \int_{B(x,\hat{r}^m)\setminus B(x,r)} |k(x,y)|\,|\chi_F(y)|\, d\mu(y)$$

$$+ \sum_{j=1}^k \int_{B(x,r_j^m)\setminus B(x,r_{j-1}^m)} \dots .$$

The first term vanishes if $\hat{r}^m > \mathcal{R}(x)$. Otherwise it is bounded by

$$\frac{1}{\mathcal{R}(x)^m}\int_{B(x,\hat{r}^m)}|\chi_F(y)|\, d\mu(y) = \frac{1}{\hat{r}^m{}^2}\int_{B(x,\hat{r}^m)}|\chi_F(y)|\, d\mu(y) \leq 2M \,.$$

Similarly

$$\int_{B(x,r_j^m)\setminus B(x,r_{j-1}^m)} |k(x,y)|\,|\chi_F(y)|\,d\mu(y) \le \frac{\mu_j}{r_{j-1}^m}\,.$$

But we know that $r_{j-1}^m = 3^{-k+j-1}r_k^m$, $\mu_j \le 6(6)^{-k+j}\mu_k$. Hence

$$\frac{\mu_j}{r_{j-1}^m} \le 6\cdot 3^{k-j+1}6^{-k+j}\frac{\mu_k}{r_k^m} \le AM2^{-k+j}\,.$$

Therefore,

$$\sum_{j=1}^{k}\int_{B(x,r_j^m)\setminus B(x,r_{j-1}^m)} |k(x,y)|\,|\chi_F(y)|\,d\mu(y) \le AM\sum_{j=1}^{k}2^{-k+j} \le AM\,.$$

We finally get

$$|(T^r f)(x) - (T^{3R}f)(x)| \le AM\,.$$

Now we need to estimate $(T^{3R}\chi_F)(x)$. Consider the average

$$V_R(x) := \frac{1}{\mu(B(x,R))}\int_{B(x,R)} T\chi_F\,d\mu\,.$$

Firstly, by the choice of R, we have

$$|V_R(x)| \le \frac{\mu(B(x,3R))}{\mu(B(x,R))}M_\mu[T\chi_F](x) \le 6M_\mu[T\chi_F](x)\,.$$

Second,

$$V_R(x) - (T^{3R}f)(x) = \int_{\mathbb{R}^n\setminus B(x,3R)} T'[\delta_x - \frac{1}{\mu(B(x,R))}\chi_{B(x,R)}\,d\mu]\chi_F\,d\mu$$

$$-\frac{1}{\mu(B(x,R))}\int_{B(x,R)} T[\chi_{F\cap B(x,3R)}]\,d\mu = I + II\,.$$

Here T' denotes the operator with kernel $k(y,x)$.

Estimate of I. Put $\eta = \delta_x - \frac{1}{\mu(B(x,R))}\chi_{B(x,R)}\,d\mu$. All radii greater than $3R$ are M-Ahlfors for μ. This and the fact that $\eta(\mathbb{C}) = 0$ allows us to use the Calderón–Zygmund property of $k(y,x)$ to prove as usual (see Lemma 11.4 below for example) that $I \le A\,\|\eta\|\,\sup_{\rho\ge R}\frac{\mu(B(x,\rho))}{\rho^m} \le AM$.

Estimate of II.

$$|II| \le \frac{1}{\mu(B(x,R))}\|\chi_{B(x,R)}\|_{L^2(\mu)}\|T(\chi_{F\cap B(x,3R)})\|_{L^2(\mu)}$$

$$\le \|T\|_2\frac{(\int_{B(x,3R)}|\chi_F|^2\,d\mu)^{\frac12}}{\mu(B(x,R))^{\frac12}}\,.$$

We can continue as follows:

$$|II| \le \|T\|_\beta\frac{\mu(B(x,3R))^{\frac12}}{\mu(B(x,R))^{\frac12}} \le 6^{\frac12}\|T\|_2 \le A\|T\|_2\,.$$

The lemma is completely proved. $\square$

Now we are ready to repeat the considerations of Theorem 5.1 of [**44**] (with small modifications due to the fact that μ is arbitrary and may not be of order m).

We are going to prove now that

$$\|T\|_{L^1(\mu)\to L^{1,\infty}} \leq A_1 CM + A_2 C\|T\|_2\,,$$

where C depend only on Calderón–Zygmund constants of the kernel of T.

Let $\nu \in M(\mathbb{C})$ be a finite linear combination of unit point masses with positive coefficients, i.e.,

$$\nu = \sum_{i=1}^{N} \alpha_i \delta_{x_i}.$$

THEOREM 11.3. *Let k satisfy* (11.1). *Then*

$$\|T\nu\|_{L^{1,\infty}(\mu)} \leq (A_1 CM + A_2 C\|T\|_2)\|\nu\|\,,$$

where C depends on n, m, M, and Calderón–Zygmund constants of the kernel.

Remark. 1) There is no problem with the definition of $T\nu$: it is just the finite sum $\sum_{i=1}^{N} \alpha_i K(x, x_i)$, which makes sense everywhere except at finitely many points.

2) The next proof *is not* via Calderón–Zygmund decomposition.

Proof. In this proof $B(x, \rho)$ denotes a closed ball, $B'(x, \rho)$ denotes an open ball. Without loss of generality, we may assume that $\|\nu\| = \sum_i \alpha_i = 1$ (this is just a matter of normalization), and of course all our α's are strictly positive. Thus we have to prove that

$$\|T\nu\|_{L^{1,\infty}(\mu)} \leq A_4\,.$$

Fix some $t > 0$ and suppose first that $\mu(\mathbb{C}) > \frac{1}{t}$. Let $B(x_1, \rho_1)$ be the smallest (closed) ball such that $\mu(B(x_1, \rho_1)) \geq \frac{\alpha_1}{t}$ (since the function $\rho \to \mu(B(x_1, \rho))$ is increasing and continuous from the right, tends to 0 as $\rho \to 0$, and is greater than $\frac{1}{t} > \frac{\alpha_1}{t}$ for sufficiently large $\rho > 0$, such ρ_1 exists and is strictly positive).

Note that for the corresponding *open* ball $B'(x_1, \rho_1) := \{y \in \mathbb{C} : \operatorname{dist}(x_1, y) < \rho_1\}$, we have $\mu(B'(x_1, \rho_1)) = \lim_{\rho \to \rho_1 - 0} \mu(B(x_1, \rho)) \leq \frac{\alpha_1}{t}$. Since the measure μ is σ-finite and non-atomic, one can choose a Borel set E_1 satisfying

$$B'(x_1, \rho_1) \subset E_1 \subset B(x_1, \rho_1) \quad \text{and} \quad \mu(E_1) = \frac{\alpha_1}{t}.$$

Let $B(x_2, \rho_2)$ be the smallest ball such that $\mu(B(x_2, \rho_2) \setminus E_1) \geq \frac{\alpha_2}{t}$ (since $\mu(\mathbb{C}) > \frac{1}{t}$, the measure of the remaining part $\mathbb{R}^n \setminus E_1$ is still greater than $\frac{1 - \alpha_1}{t} > \frac{\alpha_2}{t}$). Again for the corresponding open ball $B'(x_2, \rho_2)$, we have $\mu(B'(x_2, \rho_2) \setminus E_1) \leq \frac{\alpha_2}{t}$, and therefore there exists a Borel set E_2 satisfying

$$B'(x_2, \rho_2) \setminus E_1 \subset E_2 \subset B(x_2, \rho_2) \setminus E_1 \quad \text{and} \quad \mu(E_2) = \frac{\alpha_2}{t}.$$

In general, for $i = 3, 4, \ldots, N$, let $B(x_i, \rho_i)$ be the smallest ball such that

$$\mu\Big(B(x_i, \rho_i) \setminus \bigcup_{\ell=1}^{i-1} E_\ell\Big) \geq \frac{\alpha_i}{t},$$

and let E_i be a Borel set satisfying

$$B'(x_i, \rho_i) \setminus \bigcup_{\ell=1}^{i-1} E_\ell \subset E_i \subset B(x_i, \rho_i) \setminus \bigcup_{\ell=1}^{i-1} E_\ell \quad \text{and} \quad \mu(E_i) = \frac{\alpha_i}{t}.$$

Put $E := \bigcup_i E_i$. Clearly

$$\bigcup_i B'(x_i, \rho_i) \subset E \subset \bigcup_i B(x_i, \rho_i) \quad \text{and} \quad \mu(E) = \frac{1}{t}.$$

Now let us compare $T\nu$ to $t \sum_i \chi_{\mathbb{R}^n \setminus B(x_i, 2\rho_i)} \cdot T\chi_{E_i} =: t\sigma$ outside E. We have

$$T\nu - t\sigma = \sum_i \varphi_i$$

where

$$\varphi_i = \alpha_i T\delta_{x_i} - t \chi_{\mathbb{R}^n \setminus B(x_i, 2\rho_i)} \cdot T\chi_{E_i}.$$

Note now that

$$\int_{\mathbb{R}^n \setminus E} |\varphi_i| d\mu \leq \int_{\mathbb{R}^n \setminus B(x_i, 2\rho_i)} \left| T[\alpha_i \delta_{x_i} - t\chi_{E_i} d\mu] \right| d\mu$$

$$+ \int_{B(x_i, 2\rho_i) \setminus B'(x_i, \rho_i)} \alpha_i |T\delta_{x_i}| d\mu =: I + \alpha_i II.$$

To estimate I, notice that it has the form $\int_{\mathbb{R}^n \setminus B(x, 2\rho)} |T\eta| \, d\mu$ with the measure η supported by $B(x, \rho)$ and $\eta(\mathbb{C}) = 0$. To estimate such an integral we put $\hat{r}^m := \max[2\rho, R(x)]$ and split

$$\int_{\mathbb{R}^n \setminus B(x, 2\rho)} |T\eta| \, d\mu = \int_{B(x, \hat{r}^m) \setminus B(x, 2\rho)} |T\eta| \, d\mu + \int_{\mathbb{R}^n \setminus B(x, \hat{r}^m)} |T\eta| \, d\mu =: I_1 + I_2.$$

The integral I_2 is estimated exactly as in Lemma 11.4 because our measure is $2M$-Ahlfors for disks centered at x with radii larger than $\hat{r}^m$. Hence $I_2 \leq ACM\|\eta\| \leq ACM\alpha_i$. On the other hand, using the properties of the kernel of T, we see that

$$I_1 \leq C \min\left[\frac{1}{(2\rho)^m}, \frac{1}{R(x)^m}\right] \mu(B(x, \hat{r}^m)) \|\eta\| \leq ACM\alpha_i.$$

Hence $I \leq ACM\alpha_i$.

To estimate II we notice that it has the form $\int_{B(x, 2\rho) \setminus B(x, \rho)} |T\delta_x| \, d\mu$. This is almost the same as I_1. Namely,

$$II \leq AC \min\left[\frac{1}{\rho^m}, \frac{1}{R(x)^m}\right] \mu(B(x, 2\rho)) \leq \frac{AC\mu(B(x, 2\max[R(x), \rho]))}{\max[R(x), \rho]^m}.$$

This is bounded by ACM because our measure is $2M$-Ahlfors for disks centered at x with radii larger than $R(x)$. Finally $I + \alpha_i II \leq ACM\alpha_i$.

Finally we conclude that

$$\int_{\mathbb{R}^n \setminus E} |T\nu - t\sigma| d\mu \leq ACM \sum_i \alpha_i = ACM,$$

and thereby $|T\nu - t\sigma| \leq ACMt$ everywhere on $\mathbb{R}^n \setminus E$, except, maybe, a set of measure $\frac{1}{t}$. To accomplish the proof of the theorem, we will show that for sufficiently large $B = B(C, M, \|T\|_2)$,

$$\mu\{|\sigma| > B\} \leq \frac{2}{t}.$$

Then, combining all the above estimates, we shall get

$$\mu\{x \in \mathbb{C} \;:\; |T\nu(x)| > (B + ACM)t\} \le \frac{4}{t}.$$

We will apply the Stein-Weiss duality trick. Assume that the inverse inequality $\mu\{|\sigma| > B\} > \frac{2}{t}$ holds. Then either $\mu\{\sigma > B\} > \frac{1}{t}$, or $\mu\{\sigma < -B\} > \frac{1}{t}$. Assume for definiteness that the first case takes place and choose some set $F \subset \mathbb{C}$ of measure exactly $\frac{1}{t}$ such that $\sigma > B$ everywhere on F. Then, clearly,

$$\int_{\mathbb{C}} \sigma \chi_F \, d\mu > \frac{B}{t}.$$

On the other hand, this integral can be computed as

$$\sum_i \int_{\mathbb{C}} [T\chi_{E_i}] \cdot \chi_{F \setminus B(x_i, 2\rho_i)} \, d\mu = \sum_i \int_{\mathbb{C}} \chi_{E_i} \cdot [T' \chi_{F \setminus B(x_i, 2\rho_i)}] \, d\mu.$$

Fix a point $x \in E_i \subset B(x_i, \rho_i)$. We will use again the property that $|K(x,y)| \le \frac{1}{R(x)^m}$:

$$|T' \chi_{F \setminus B(x_i, 2\rho_i)}(x) - T' \chi_{F \setminus B(x, \rho_i)}(x)|$$

$$\le \int_{B(x_i, 2\rho_i) \setminus B(x, \rho_i)} |K(y,x)| \, d\mu(y) \le \frac{AC\mu(B(x, 3\max[\rho_i, R(x)]))}{\max[\rho_i, R(x)]^m} \le ACM,$$

because all disks centered at x and of radii greater than $R(x)$ are $2M$-Ahlfors, and therefore for every $x \in E_i \cap \operatorname{supp}\mu$,

$$|T' \chi_{F \setminus B(x_i, 2\rho_i)}(x)| \le (T')^* \chi_F(x) + ACM \le 2 \cdot A \, M_\mu T' \chi_F(x) + ACM$$

according to Lemma 11.2. Hence

$$\int_{\mathbb{C}} \sigma \chi_F \, d\mu \le ACM\mu(E) + 2 \cdot A \int_{\mathbb{C}} \chi_E \cdot M_\mu T' \chi_F \, d\mu.$$

But the first term equals $\dfrac{ACM}{t}$ while the second one does not exceed

$$2 \cdot A \, \|\chi_E\|_{L^2(\mu)} \|M_\mu T' \chi_F\|_{L^2(\mu)} \le \frac{2 \cdot A}{t} \|M_\mu\|_{L^2(\mu) \to L^2(\mu)} \|T'\|_{L^2(\mu) \to L^2(\mu)}.$$

Recalling that $\|T'\|_{L^2(\mu) \to L^2(\mu)} = \|T\|_{L^2(\mu) \to L^2(\mu)}$, we see that one can take

$$B = ACM + 2 \cdot A \, \|M_\mu\|_{L^2(\mu) \to L^2(\mu)} \|T\|_{L^2(\mu) \to L^2(\mu)}$$

to get a contradiction. Since the norm $\|M_\mu\|_{L^2(\mu) \to L^2(\mu)}$ is bounded by some absolute constant, we are done. $\qquad\square$

Now we are ready to pass from weak type estimates for measures, which are finite linear combinations of delta measures, to the weak type estimate for arbitrary $L^1(\mu)$ functions.

We shall need one more standard observation about generalized Calderón–Zygmund kernels satisfying (11.1). We used it already; now let us prove it.

LEMMA 11.4. *Let k satisfy* (11.1). *Let $\eta \in M(\mathcal{X})$, $\eta(\mathcal{X}) = 0$, and $\operatorname{supp}\eta \subset B(x, \rho)$ for some $\rho > 0$. Then for every non-negative Borel measure ν on $\mathcal{X}$, we have*

$$\int_{\mathcal{X} \setminus B(x, 2\rho)} |T\eta| \, d\nu \le A_1 \, M_\mu \nu(x) \, \|\eta\|,$$

where $A_1 > 0$ depends only on the dimension n and the constants A and τ in the definition of the Calderón–Zygmund kernel K. In particular,

$$\int_{\mathcal{X} \setminus B(x,2\rho)} |T\eta| \cdot |f| d\mu \leq A_1 \, M_\mu f(x) \, \|\eta\|$$

for every Borel measurable function f on $\mathcal{X}$, and

$$\int_{\mathcal{X} \setminus B(x,2\rho)} |T\eta| d\mu \leq A_1 \, \|\eta\| .$$

Proof. Let x_0 be the point in $B(x,\rho)$ such that $\mathcal{R}(x_0) \leq 2 \inf_{y \in B(x,\rho)} \mathcal{R}(y)$. If $\mathcal{R}(x_0) \leq \rho$, then $\mu(B(x_0,s)) \leq 2M \, s^m, s \geq 2\rho$. Also we notice that

$$|T\eta(y)| = \left| \int_{B(x,\rho)} k(y,x') d\eta(x') \right| = \left| \int_{B(x,\rho)} [k(y,x') - k(y,x_0)] d\eta(x') \right| \leq$$

$$\leq \|\eta\| \sup_{x' \in B(x,\rho)} |k(y,x') - k(y,x_0)| \leq \|\eta\| \frac{A\rho^\tau}{\text{dist}(x_0,y)^{m+\tau}} .$$

Then the usual estimate (here $x_0 \in B(x,\rho)$)

$$\int_{y:|y-x| \geq 2\rho} \frac{\rho^\tau}{\text{dist}(x_0,y)^{m+\tau}} \, d\mu(y) \leq \int_{y:|y-x| \geq 2\rho} \frac{A\rho^\tau}{\text{dist}(x,y)^{m+\tau}} \, d\mu(y) \leq CM_\mu \nu(x)$$

finishes the lemma. The last inequality follows from the simple observation that for every $r \geq 2\rho$, we have $\mu(B(x,3r)) \leq \mu(B(x_0,3.5r)) \leq (3.5)^m M r^m$ because $\mathcal{R}(x_0) \leq \rho$.

Let $\mathcal{R}(x_0) \geq 20\,\rho$. In this case we split the domain $\{y : |y-x| \geq 2\rho\}$ into two domains: $I := \{y : |y-x| \leq 2\mathcal{R}(x_0)\}$, $II := \{y : |y-x| \geq 2\mathcal{R}(x_0)\}$. The integral over the second domain can be estimated by subtracting from the kernel $k(y,x')$ the number $k(y,x_0)$ and repeating verbatim the first part of the proof. On the other hand,

$$\int_I \int_{B(x,\rho)} |k(y,x')| d|\eta|(x') d\nu(y) \leq \frac{C \, \nu(B(x,2\mathcal{R}(x_0))}{\mathcal{R}(x_0)^m} \|\eta\| .$$

At the same time $\mu(B(x,6\mathcal{R}(x_0))) \leq \mu(B(x_0,6.3\mathcal{R}(x_0))) \leq (6.3)^m M\mathcal{R}(x_0)^m$. So we can continue the previous estimate as follows:

$$\int_I \int_{B(x,\rho)} |k(y,x')| d|\eta|(x') d\nu(y) \leq C(m,M)\|\eta\| M_\mu \nu(x) .$$

It remains to consider the possibility that $\rho \leq \mathcal{R}(x_0) \leq 20\,\rho$. In this case consider $I := \{y : |y-x| \leq 20\rho\}$, $II := \{y : |y-x| \geq 20\rho\}$ splitting the domain $\{y : |y-x| \geq 2\rho\}$. The integral II is taken care of exactly as before. On the other hand,

$$\int_I \int_{B(x,\rho)} |k(y,x')| d|\eta|(x') d\nu(y) \leq \frac{C \, \nu(B(x,20\rho)}{\rho^m} \|\eta\| .$$

At the same time $\mu(B(x,60\rho)) \leq \mu(B(x_0,61\rho)) \leq 61^m M\rho^m$ because $61\rho > \mathcal{R}(x_0)$. So we can continue the previous estimate like that:

$$\int_I \int_{B(x,\rho)} |k(y,x')| d|\eta|(x') d\nu(y) \leq C(m,M)\|\eta\| M_\mu \nu(x) .$$

$\square$

THEOREM 11.5. *Let k satisfy (11.1). Then for an arbitrary measure ν,*

$$\|T\nu\|_{L^{1,\infty}(\mu)} \leq (A_1 CM + A_2 C\|T\|_2)\|\nu\|,$$

where C depends on n, m, M, and Calderón–Zygmund constants of the kernel.

Note first of all that Theorem 11.3 remains valid (with constant doubled) for finite linear combinations of point masses with *arbitrary* real coefficients. Indeed, every such measure ν can be represented as $\nu_+ - \nu_-$ where $\nu_\pm$ are finite linear combinations of point masses with *positive* coefficients and $\|\nu\| = \|\nu_+\| + \|\nu_-\|$. Hence

$$\begin{aligned}
\|T\nu\|_{L^{1,\infty}(\mu)} &\leq 2\big(\|T\nu_+\|_{L^{1,\infty}(\mu)} + \|T\nu_-\|_{L^{1,\infty}(\mu)}\big) \\
&\leq 2A_4(\|\nu_+\| + \|\nu_-\|) = 2A_4\|\nu\|.
\end{aligned}$$

Proof. Let $C_0(\mathcal{X})$ be the space of bounded continuous functions on the space $\mathcal{X}$ with bounded support (a function is said to have bounded support if it vanishes outside some (large) ball of finite radius). Clearly, $C_0(\mathcal{X}) \subset L^1(\mu) \cap L^2(\mu)$, and it is a standard fact from measure theory that $C_0(\mathcal{X})$ is dense on $L^1(\mu) \cap L^2(\mu)$ with respect to the norm $\|\cdot\|_{L^1(\mu)} + \|\cdot\|_{L^2(\mu)}$. Therefore it is enough to prove the desired inequality for $f \in C_0(\mathcal{X})$.

Fix $t > 0$ and put $G := \{x \in \mathcal{X} : |f(x)| > t\}$, $f^t := f \cdot \chi_G$, and $f_t = f \cdot \chi_{\mathcal{X} \setminus G}$. We have $Tf = Tf^t + Tf_t$. Now observe, as usual, that

$$\int_{\mathcal{X}} |f_t|^2 \, d\mu \leq t \int_{\mathcal{X}} |f_t| \, d\mu \leq t\|f\|_{L^1(\mu)}.$$

Therefore $\int_{\mathcal{X}} |Tf_t|^2 \, d\mu \leq \|T\|^2_{L^2(\mu) \to L^2(\mu)} t\|f\|_{L^1(\mu)}$, and

$$\mu\big\{x \in \mathcal{X} : |Tf_t(x)| > t \cdot \|T\|_{L^2(\mu) \to L^2(\mu)}\big\} \leq \frac{\|f\|_{L^1(\mu)}}{t}.$$

Note now that G is an *open* set (this is the only place where we use the continuity of f) and that $\mu(G) \leq \frac{1}{t}\|f\|_{L^1(\mu)}$. Recall that every open set G in a separable metric space allows a "Whitney decomposition"; i.e., it can be represented as a union of countably many pairwise disjoint Borel sets G_i $(i = 1, 2, \dots)$ satisfying

$$\operatorname{diam} G_i \leq \tfrac{1}{2} \operatorname{dist}(G_i, \mathcal{X} \setminus G).$$

Put $f_i := f \cdot \chi_{G_i}$. Then $f^t = \sum_{i=1}^{\infty} f_i$ where the series converges at least on $L^2(\mu)$. Let $f^{(N)}$ be the N-th partial sum of this series. Define

$$\alpha_i := \int_{\mathcal{X}} f_i \, d\mu = \int_{G_i} f \, d\mu.$$

Obviously, $\sum_{i=1}^{\infty} |\alpha_i| \leq \|f\|_{L^1(\mu)}$. Choose one point x_i in every set G_i and put $\nu_N = \sum_{i=1}^{N} \alpha_i \delta_{x_i}$. Consider the difference $Tf^{(N)} - T\nu_N$ outside G. We have

$$\int_{\mathcal{X} \setminus G} |Tf^{(N)} - T\nu_N| \, d\mu \leq \sum_{i=1}^{N} \int_{\mathcal{X} \setminus G} |T[f_i d\mu - \alpha_i \delta_{x_i}]| \, d\mu \leq 2A_1 \sum_{i=1}^{N} |\alpha_i| \leq 2A_1\|f\|_{L^1(\mu)}$$

according to the standard Calderón–Zygmund estimate, but using (11.1). See Lemma 11.4. Thus $|Tf^{(N)} - T\nu_N| \leq 2A_1 t$ everywhere outside G save, maybe, some exceptional set of measure at most $\frac{1}{t}\|f\|_{L^1(\mu)}$. As we have seen above,

$$\mu\{x \in \mathcal{X} : |T\nu_N(x)| > 2A_4 t\} \le \frac{1}{t}\|\nu_N\| \le \frac{1}{t}\|f\|_{L^1(\mu)}.$$

Hence

$$\mu\{x \in \mathcal{X} \setminus G : |Tf^{(N)}(x)| > 2(A_1 + A_4)t\} \le \frac{2}{t}\|f\|_{L^1(\mu)},$$

and

$$\mu\{x \in \mathcal{X} : |Tf^{(N)}(x)| > 2(A_1 + A_4)t\} \le \frac{3}{t}\|f\|_{L^1(\mu)}.$$

Since $f^{(N)} \to f^t$ on $L^2(\mu)$ as $N \to +\infty$, we have $Tf^{(N)} \to Tf^t$ on $L^2(\mu)$ as $N \to +\infty$, which is more than enough to pass to the limit and to conclude that

$$\mu\{x \in \mathcal{X} : |Tf^t(x)| > 2(A_1 + A_4)t\} \le \frac{3}{t}\|f\|_{L^1(\mu)}.$$

Thus, we can take $A_5 = 4\big[\|T\|_{L^2(\mu)\to L^2(\mu)} + 2(A_1 + A_4)\big]$.   □

The proof of Theorem 11.5 is finished, and so is the proof of Theorem 11.1 because it follows from the weak type estimate for T and the Lemma of David (see the proof of Theorem 7.1 on pages 479–480 of [**44**]).   □

CHAPTER 12

Starting the Proof of Nonhomogeneous Nonaccretive Tb Theorem

We are going to prove Theorem 10.6. Let us also recall its statement. The proof will be completely independent of what has been written above. This proof is the backbone of what has been done, in particular, of semiadditivity of Lipschitz harmonic capacity (our main result). But it has other applications, which will be explained elsewhere.

Let us fix $\omega = (\omega_1, \omega_2) \in \Omega^2$; this is the same as to fix two dyadic lattices $\mathcal{D}_1$ and $\mathcal{D}_2$ satisfying (7.19).

Let η be finite and positive constant. A cube Q of any of these lattices we call η non-accretive if

$$(12.1) \qquad |\nu(Q)| = |\int_Q b \, d\mu| \leq \eta \mu(Q) \,.$$

Notice that cube Q with $\mu(Q) = 0$ is η non-accretive.

Cubes are open. Set $T_{\mathcal{D}_1}, T_{\mathcal{D}_2}$ to be unions of all η non-accretive cubes from $\mathcal{D}_1$ and $\mathcal{D}_2$ correspondingly. Let $T_{12} := T_{\mathcal{D}_1} \cup T_{\mathcal{D}_2}$. We call T_{12} the *non-accretive set* with respect to this pair of lattices. The reader should think that this open set is not too big. Let us repeat the statement of Theorem 10.6.

THEOREM 12.1. *Let k be a continuous Calderón–Zygmund kernel with constant C and of order m. Let b satisfy (7.5) with constant $\hat{A}$. Let*

$$(12.2) \qquad T_* b \leq C_6 \,.$$

Let k satisfy

$$(12.3) \qquad |k(x,y)| \leq \min[\frac{1}{R(x)^m}, \frac{1}{R(y)^m}] \,,$$

where $R(x)$ is the supremum of radii of all B non-Ahlfors balls centered at x. Let k satisfy

$$(12.4) \qquad |k(x,y)| \leq \min[\frac{1}{d(x)^m}, \frac{1}{d(y)^m}] \,,$$

where $d(x) := \operatorname{dist}(x, \mathbb{R}^n \setminus T_{12})$. There exists C_8 depending only on $B, C, m, n, C_6, \hat{A}$ (and not depending on T, f, g) such that for every positive δ,

$$(12.5) \qquad |\langle Tf, g \rangle| \leq A \, C_8 \, \eta^{-4} \delta^{-A} \|f\| \|g\| + \|T\| R(\omega, f, g) \,,$$

where $R(\omega, f, g)$ does not depend on T, and we have the following estimate on the mathematical expectation of the remainder $R(\omega, f, g)$:

$$(12.6) \qquad \mathbb{E} R(\omega, f, g) \leq A \delta \|f\| \|g\| \,.$$

Remark. Measure μ is *arbitrary* here–no restrictions on μ, any growth, any decay is allowed, no doubling of course is assumed. The Calderón–Zygmund kernels with (12.3) are adapted to this measure in the sense that $|k(x,y)|$ tries to be "not too large" if $y \in B(x, R)$ with (relative to R^m) very large measure. We will see later that this is a very natural restriction on k, μ for *arbitrary* μ.

12.1. Terminal and transit cubes

Given a constant D, consider

$$M = M(D) := \{x \in F : \text{every ball centered at } x \text{ is } D\text{Ahlfors}\}.$$

Any such set M is closed. We use the assumption $m \leq n - 1$ in the following lemma.

LEMMA 12.2. *There exists $A < \infty$ depending only on n such that*

$$(12.7) \qquad\qquad R(x) \geq \text{dist}(x, M(AB)),$$

if $R(x)$ denote the supremum of radii of all B non-Ahlfors balls.

Proof. Let us show that at least one point of the sphere $\partial B(x, R(x))$ belongs to $M(AB)$ if A is sufficiently large. Fix $y \in \partial B(x, R(x))$, the estimate $\mu(B(y, R)) \leq 2^m BR^m$ is obvious for $R \geq R(x)$ just because $B(y, R) \subset B(x, 2R)$, and this latter ball *is* B Ahlfors. So if we assume that $\partial B(x, R(x))$ does not lie in $M(AB)$, then

$$\forall y \in \partial B(x, R(x)) \, \exists r(y) < R(x) : \mu(B(y, r)) > AB \, r(y)^m.$$

$\{B(y, r(y))\}_{y \in \partial B(x, R(x))}$ covers $\partial B(x, R(x))$, and let $\{B_i\} = \{B(y_i, r(y_i))\}$ be a disjoint subcover such that

$$\partial B(x, R(x)) \subset \cup_i 3B_i.$$

Then

$$\mu(B(x, 2R(x))) \geq \sum_i \mu(B_i) \geq AB \sum_i r(y_i)^m \geq AB3^{-m}\left(\sum_i (3R(y_i))^{n-1}\right)^{\frac{m}{n-1}}.$$

Here $m \leq n - 1$ was used.

Set

$$S_i := 3B_i \cap \partial B(x, R(x)).$$

There exists a constant $b(n)$ such that $(3r(y_i))^{n-1} \geq area(S_i)$. We can now continue our inequality:

$$\left(\sum_i (3R(y_i))^{n-1}\right)^{\frac{m}{n-1}} \geq b(n)^{\frac{m}{n-1}}\left(\sum area(S_i)\right)^{\frac{m}{n-1}}$$

$$\geq b(n)^{\frac{m}{n-1}} area(\partial B(x, R(x)))^{\frac{m}{n-1}}.$$

But the last expression is obviously larger than $a(n)R(x)^m$. Putting all together, we see that

$$\mu(B(x, 2R(x))) \geq 10^{-m} AB \, a(n)(2R(x))^m.$$

It is sufficient to choose $A = 11^m a(n)^{-1}$ to come to the contradiction. Lemma 12.2 is proved. $\qquad\square$

So let us look at (12.3), take B that defines $R(x)$ in it, take A from Lemma 12.2, and set

$$(12.8) \qquad\qquad H := \mathbb{R}^n \setminus M(AB).$$

We call the cube Q from $\mathcal{D}_1$ a *terminal cube*, if $2Q$ is contained in H or Q is contained in a cube of a family $T_{\mathcal{D}_1}$ (so it is η non-accretive). The same with cubes from $\mathcal{D}_2$: we call the cube R from $\mathcal{D}_2$ a *terminal cube*, if $2R$ is contained in H or R is contained in a cube of the family $T_{\mathcal{D}_2}$ (so it is η non-accretive). All other cubes from $\mathcal{D}_1$ and $\mathcal{D}_2$ are called *transit cubes*.

It will be very important on terminal cubes that the kernel is very well bounded from above on these cubes. Let us denote by $\mathcal{D}_1^{term}$, $\mathcal{D}_1^{tr}$ the terminal and transit cubes from $\mathcal{D}_1$. Symbols $\mathcal{D}_2^{term}$, $\mathcal{D}_2^{tr}$ denote the terminal and transit cubes from $\mathcal{D}_2$.

LEMMA 12.3. *Let Q belong to $\mathcal{D}_1^{term}$; then*

$$(12.9) \qquad |k(x,y)| \leq \frac{1}{\operatorname{dist}(x,\partial Q)^m} \; \forall x \in Q.$$

If Q belongs to $\mathcal{D}_1^{term}$ because $2Q \subset H$, then

$$(12.10) \qquad |k(x,y)| \leq \frac{1}{\ell(Q)^m} \; \forall x \in Q.$$

In particular, the latter is the case if all cubes are η accretive. The same statements hold for R's from $\mathcal{D}_2^{term}$.

Proof. If Q is contained in $T_{\mathcal{D}_1}$ (is η non-accretive) then we use (12.4):

$$|k(x,y)| \leq \min\left[\frac{1}{d(x)^m}, \frac{1}{d(y)^m}\right],$$

where $d(x) := \operatorname{dist}(x, \mathbb{R}^n \setminus T_{12})$. We then notice that $Q \subset T_{\mathcal{D}_1} \subset T_{12}$ imply that $d(x) \geq \operatorname{dist}(x, \partial Q)$. If $2Q \subset H$, then we use (12.3) and Lemma 12.2: $R(x) \geq \operatorname{dist}(x, H) \geq \operatorname{dist}(x, \partial 2Q) \geq \ell(Q)$. We are done. $\qquad \square$

LEMMA 12.4. *There exists $A_0 < \infty$ which depends only on n such that if Q belongs to $\mathcal{D}_1^{tr}$, then*

$$(12.11) \qquad \mu(Q) \leq A_0 B \ell(Q)^m.$$

The same holds for R from $\mathcal{D}_2^{tr}$.

Proof. In fact, if Q is in $\mathcal{D}_1^{tr}$, then $2Q$ necessarily intersects ∂H. Let x be in the intersection. Then there exists $A(n)$ such that $Q \subset B(x, A(n)\ell(Q))$. But all balls centered at ∂H are AB Ahlfors, so $\mu(Q) \leq A(n)^m B \ell(Q)^m$, and we are done. $\qquad \square$

There are special unit cubes: the unit cube Q^0 of $\mathcal{D}_1$, which contains F and the unit cube R^0 of $\mathcal{D}_2$, which contains F. Notice that by (7.19) both the unit cube Q^0 and the unit cube R^0 contain F deep inside them. They are terminal simultaneously, and this happens if and only if either $M \cap F = \emptyset$ or

$$(12.12) \qquad |\int_F b\,d\mu| \leq \eta\|\mu\|.$$

But then Lemma 12.3 gives obviously that $|k(x,y)| \leq 10^m$ for all $x, y \in F$ (we use (12.9) and the fact that F lies deep inside Q^0). Then of course $\langle Tf, g\rangle| \leq C(m)\|f\|\|g\|$ just because of the estimate on the kernel: $|k(x,y)| \leq 10^m$.

So we can think that Q^0, R^0 are transit cubes.

12.2. Projections Λ and Δ_Q

When we introduced the notion of good functions and bad functions, we already introduced these projections. Here we repeat.

Let $\mathcal{D}_1$ be one of the dyadic lattices above. For a function $\psi \in L^1(\mu)$ and for a cube $Q \subset \mathbb{C}$, denote by $\langle \psi \rangle_Q$ the average value of ψ over Q with respect to the measure μ, i.e.,

$$\langle \psi \rangle_Q := \frac{1}{\mu(Q)} \int_Q \psi \, d\mu$$

(of course, $\langle \psi \rangle_Q$ makes sense only for cubes Q with $\mu(Q) > 0$).

Put

$$\Lambda \varphi := \frac{\langle \varphi \rangle_{Q^0}}{\langle b \rangle_{Q^0}} b.$$

Clearly, $\Lambda \varphi \in L^2(\mu)$ for all $\varphi \in L^2(\mu)$, and $\Lambda^2 = \Lambda$, i.e., Λ is a projection. Note also, that actually Λ does not depend on the lattice $\mathcal{D}_1, \mathcal{D}_2$, because the average is taken over the whole support of the measure μ regardless of the position of the cube Q^0 (or R^0).

From now on, we will always denote by Q_j ($j = 1, \ldots, 2^n$) the 2^n dyadic subcubes of a cube Q enumerated in some "natural order". In particular, that means that we will have to give up our idea to denote the cubes in $\mathcal{D}_1$ and $\mathcal{D}_2$ by Q_1 and Q_2, respectively. So we use the letter R with subscripts and superscripts to denote the cubes from $\mathcal{D}_2$. In particular, we will always denote by R_j ($j = 1, \ldots, 2^n$) the 2^n dyadic subcubes of a cube R from $\mathcal{D}_2$. Below we will start almost every claim by "Assume (for definiteness) that $\ell(Q) \leq \ell(R)$ ... ".

For every transit cube $Q \in \mathcal{D}_1$, define $\Delta_Q \varphi$ by

$$\Delta_Q \varphi \big|_{\mathbb{R}^n \setminus Q} := 0, \qquad \Delta_Q \varphi \big|_{Q_j} := \begin{cases} \left[\dfrac{\langle \varphi \rangle_{Q_j}}{\langle b \rangle_{Q_j}} - \dfrac{\langle \varphi \rangle_Q}{\langle b \rangle_Q} \right] b & \text{if } Q_j \text{ is transit;} \\[4mm] \varphi - \dfrac{\langle \varphi \rangle_Q}{\langle b \rangle_Q} b & \text{if } Q_j \text{ is terminal} \end{cases}$$

($j = 1, \ldots, 2^n$). Observe that for every transit cube Q, we have $\mu(Q) > 0$ and

$$|\langle b \rangle| > \eta$$

for all transit cubes, so our definition makes sense: no zero can appear in the denominator.

We repeat the same definition for $R \in \mathcal{D}_2$.

Easy properties of $\Delta_Q \varphi$. For every $\varphi \in L^2(\mu)$ and transit Q,

  1) $\Delta_Q \varphi \in L^2(\mu)$;
  2) $\int_{\mathbb{C}} \Delta_Q \varphi \, d\mu = 0$;
  3) Δ_Q is a projection, i.e., $\Delta_Q^2 = \Delta_Q$;
  4) $\Delta_Q \Lambda = \Lambda \Delta_Q = 0$;
  5) If $Q, \widetilde{Q}$ are transit, $\widetilde{Q} \neq Q$, then $\Delta_Q \Delta_{\widetilde{Q}} = 0$.

To check these properties is left to the reader as an exercise.

LEMMA 12.5. *Let Q^0 be a transit cube. For every $\varphi \in L^2(\mu)$ we have*

$$\varphi = \Lambda\varphi + \sum_{Q \, transit} \Delta_Q \varphi,$$

the series converges in $L^2(\mu)$ and, moreover,

$$a\eta^4 \|\varphi\|^2_{L^2(\mu)} \leq \|\Lambda\varphi\|^2_{L^2(\mu)} + \sum_{Q \, transit} \|\Delta_Q\varphi\|^2_{L^2(\mu)} \leq A\eta^{-4}\|\varphi\|^2_{L^2(\mu)}.$$

Proof. Note first of all that if one understands the sum

$$\sum_{Q \, transit}$$

as $\lim_{k\to\infty} \sum_{Q \, transit: \ell(Q) > 2^{-k}}$, then for μ-almost every $x \in \mathbb{C}$, one has

$$\varphi(x) = \Lambda\varphi(x) + \sum_{Q \, transit} \Delta_Q\varphi(x).$$

Indeed, the claim is obvious if the point x lies in some terminal cube. Suppose now that it is not the case. Observe that

$$\Lambda\varphi(x) + \sum_{Q \, transit: \ell(Q) > 2^{-k}} \Delta_Q\varphi(x) = \frac{\langle\varphi\rangle_{Q^k}}{\langle b\rangle_{Q^k}} b(x),$$

where Q^k is the dyadic cube of size 2^{-k}, containing x. Therefore, the claim is true if

$$\langle\varphi\rangle_{Q^k} \to \varphi(x) \quad \text{and} \quad \langle b\rangle_{Q^k} \to b(x) \quad \text{as } n \to \infty$$

(since for every transit cube Q the average $|\langle b\rangle_Q|$ is larger than η, we surely have $h(x) \neq 0$ for such x). But the exceptional set for any of these conditions has μ-measure 0.

Now let us compare $\Lambda\varphi$ and $\Delta_Q\varphi$ to the corresponding terms in the standard martingale decomposition, i.e., to

$$\widetilde{\Lambda}\varphi := \langle\varphi\rangle_{Q^0}$$

and

$$\widetilde{\Delta}_Q\varphi\big|_{\mathbb{R}^n \setminus Q} := 0, \qquad \widetilde{\Delta}_Q\varphi\big|_{Q_j} := \begin{cases} \langle\varphi\rangle_{Q_j} - \langle\varphi\rangle_Q & \text{if } Q_j \text{ is transit;} \\ \varphi - \langle\varphi\rangle_Q & \text{if } Q_j \text{ is terminal} \end{cases}$$

$(j = 1, \ldots, 2^n)$. It is well-known (and easy to prove) that

$$\|\widetilde{\Lambda}\varphi\|^2_{L^2(\mu)} + \sum_{Q \, transit} \|\widetilde{\Delta}_Q\varphi\|^2_{L^2(\mu)} = \|\varphi\|^2_{L^2(\mu)}.$$

A direct computation yields

$$\|\widetilde{\Lambda}\varphi\|^2_{L^2(\mu)} = |\langle\varphi\rangle_{Q^0}|^2 \mu(Q^0), \qquad \|\Lambda\varphi\|^2_{L^2(\mu)} = \frac{\langle|b|^2\rangle_{Q^0}}{|\langle b\rangle_{Q^0}|^2} |\langle\varphi\rangle_{Q^0}|^2 \mu(Q^0),$$

i.e.,

$$\|\Lambda\varphi\|^2_{L^2(\mu)} = \frac{\langle|b|^2\rangle_{Q^0}}{|\langle b\rangle_{Q^0}|^2} \|\widetilde{\Lambda}\varphi\|^2_{L^2(\mu)}.$$

But we have assumption (7.5) and the fact that Q^0 is transit, so the ratio

$$\frac{\langle |b|^2 \rangle_{Q^0}}{|\langle b \rangle_{Q^0}|^2} \le \frac{\hat{A}^2}{\eta^2} \,.$$

Therefore,

$$\|\widetilde{\Lambda}\varphi\|^2_{L^2(\mu)} \le \|\Lambda\varphi\|^2_{L^2(\mu)} \le \frac{\hat{A}^2}{\eta^2}\|\widetilde{\Lambda}\varphi\|^2_{L^2(\mu)} \,.$$

As to the terms $\Delta_Q\varphi$, we will represent each of them as the difference $\Delta'_Q\varphi - \frac{\langle\varphi\rangle_Q}{\langle b\rangle_Q}b_Q$, where

$$\Delta'_Q\varphi\big|_{\mathbb{R}^n\setminus Q} := 0, \qquad \Delta'_Q\varphi\big|_{Q_j} := \begin{cases} \dfrac{\langle\varphi\rangle_{Q_j} - \langle\varphi\rangle_Q}{\langle b\rangle_{Q_j}}b & \text{if } Q_j \text{ is transit;} \\[2mm] \varphi - \langle\varphi\rangle_Q & \text{if } Q_j \text{ is terminal,} \end{cases}$$

and

$$b_Q\big|_{\mathbb{R}^n\setminus Q} := 0, \qquad b_Q\big|_{Q_j} := \begin{cases} \dfrac{\langle b\rangle_{Q_j} - \langle b\rangle_Q}{\langle b\rangle_{Q_j}}b & \text{if } Q_j \text{ is transit;} \\[2mm] b - \langle b\rangle_Q & \text{if } Q_j \text{ is terminal} \end{cases}$$

$(j = 1, \ldots, 2^n)$. Note that $\Delta'\varphi \equiv \widetilde{\Delta}\varphi$ on $\mathbb{R}^n \setminus Q$ and on every terminal cube Q_j. Also, if Q_j is a transit subcube of Q, then

$$\int_{Q_j} |\widetilde{\Delta}_Q\varphi|^2 d\mu \le \int_{Q_j} |\Delta'_Q\varphi|^2 d\mu \le \frac{\hat{A}^2}{\eta^2} \int_{Q_j} |\widetilde{\Delta}_Q\varphi|^2 d\mu \,.$$

We get

$$\|\Lambda\varphi\|^2_{L^2(\mu)} + \sum_{Q\in\text{transit}} \|\Delta_Q\varphi\|^2_{L^2(\mu)} \le 2\frac{\hat{A}^2}{\eta^2}\|\varphi\|^2_{L^2(\mu)} + 2\sigma \,,$$

where

$$\sigma := \sum_{Q\in\text{transit}} \frac{|\langle\varphi\rangle_Q|^2}{|\langle b\rangle_Q|^2}\|b_Q\|^2_{L^2(\mu)} \le \frac{\hat{A}^2}{\eta^4} \sum_{Q\in\text{transit}} |\langle\varphi\rangle_Q|^2\|\widetilde{\Delta}_Q b\|^2_{L^2(\mu)} \,,$$

because $|\langle b\rangle_Q| \ge \eta$; the same reasoning we used when comparing $\Delta'_Q\varphi$ to $\widetilde{\Delta}_Q\varphi$ allows us to conclude that $\|b_Q\|^2_{L^2(\mu)} \le \frac{\hat{A}^2}{\eta^2}\|\widetilde{\Delta}_Q b\|^2_{L^2(\mu)}$.

Now let us remind the reader of the celebrated

Dyadic Carleson Imbedding Theorem.

THEOREM 12.6. *Assume that we have a dyadic lattice $\mathcal{D}$, and a family of nonnegative numbers $\{a_Q\}_{Q\in\mathcal{D}}$. Suppose also that for every cube $R \in \mathcal{D}$, we have*

$$\sum_{Q\in\mathcal{D}:Q\subset R} a_Q \le A\mu(R).$$

Then for every function $\varphi \in L^2(\mu)$, we have

$$\sum_{Q\in\mathcal{D}:\mu(Q)\ne 0} a_Q|\langle\varphi\rangle_Q|^2 \le 4A\|\varphi\|^2_{L^2(\mu)} \,.$$

The proof can be found in [**21**]. We will use this theorem now. We observe that for every transit cube $R \in \mathcal{D}$, we have

$$\sum_{Q \in \mathcal{D}^{tr}:Q \subset R} \|\widetilde{\Delta}_Q b\|^2_{L^2(\mu)} = \sum_{Q \in \mathcal{D}^{tr}:Q \subset R} \|\widetilde{\Delta}_Q (b \chi_R)\|^2_{L^2(\mu)}$$

$$\leq \|b\chi_R\|^2_{L^2(\mu)} = \int_R |b|^2 d\mu \leq \hat{A}^2 \mu(R) \,.$$

Thus, applying the Dyadic Carleson Imbedding Theorem to $a_Q = \|\widetilde{\Delta}_Q b\|^2_{L^2(\mu)}$, if Q is transit, and $a_Q = 0$, if Q is terminal, we get

$$\sigma \leq \frac{4\hat{A}^4}{\eta^4} \|f\|_{L^2(\mu)} \,.$$

Finally we proved one half of the lemma (the left inequality):

(12.13) $$\|\Lambda f\|^2 + \sum_{Q \in \text{transit}} \|\Delta_Q f\|^2 \leq \frac{A\hat{A}^2}{\eta^2} \|f\|^2 \,.$$

To finish the proof of the lemma, it remains to prove the estimate from below on $\|\Lambda \varphi\|^2_{L^2(\mu)} + \sum_{Q \in \text{transit}} \|\Delta_Q \varphi\|^2_{L^2(\mu)}$. It is sufficient to prove this estimate from below only for f's that are the finite sums of the form $\Lambda f + \sum \Delta_Q f$. In fact, for a general $f \in L^2(\mu)$, the result will then follow by a limiting argument. For finite sums $\Lambda f + \sum \Delta_Q f$ we will use the following simple lemma.

LEMMA 12.7. *Let* $P_0, P_1, \ldots, P_j, \ldots$ *be projections in the Hilbert space* $\mathcal{H}$, *and let* $P_i P_j = 0$, $i \neq j$. *Suppose also that*

(12.14) $$\sum_k \|P_j^* g\|^2 \leq C^2 \|g\|^2 \,,$$

for all $g \in \mathcal{H}$. *Suppose also that*

$$\sum_{j=0}^k P_j^* \to I \; strongly \,.$$

Then if $f = \sum_{j=0}^N P_j f$, *we have*

$$\|f\|^2 \leq C^2 \left(\sum_k \|P_j f\|^2 \right) \,.$$

Proof. Let $f = \sum_{j=0}^N P_j f$. Of course $P_m f = 0$, if $m > N$ (use $P_i P_j = 0$ assumption). Let $g \in \mathcal{H}$, $\|g\| = 1$. Let $s \geq N$. Set $g_s := \sum_{j=0}^s P_j^* g$. Then

$$|(f, g_s)| = \left| \sum_{j=0}^s (f, P_j^* g) \right| = \left| \sum_{j=0}^N (P_j f, P_j^* g) \right| \leq C \left(\sum_{j=0}^N \|P_j f\|^2 \right)^{1/2} \,.$$

This is from (12.14). As $g_s \to g$ in $\mathcal{H}$ (by the strong convergence assumption), we get $\|f\|^2 \leq C^2 (\sum_{j=0}^N \|P_j f\|^2)$. $\square$

We are going to apply this lemma to $P_0 = \Lambda$, $P_k = \Delta_Q$ (enumeration is not important). Then (12.14) holds because Δ_Q^* will satisfy the analog of inequality

(12.13):

$$(12.15) \qquad \|\Lambda^* f\|^2 + \sum_{Q \in \text{transit}} \|\Delta_Q^* f\|^2 \le \frac{A\hat{A}^2}{\eta^2} \|f\|^2.$$

To see that let us write the formula for Δ_Q^*. For every transit cube $Q \in \mathcal{D}_1$, define $\Delta_Q^* \varphi$ by

$$\Delta_Q^* \varphi \big|_{\mathbb{R}^n \setminus Q} := 0, \qquad \Delta_Q^* \varphi \big|_{Q_j} := \begin{cases} \left[\dfrac{\langle b\varphi \rangle_{Q_j}}{\langle b \rangle_{Q_j}} - \dfrac{\langle b\varphi \rangle_Q}{\langle b \rangle_Q} \right] & \text{if } Q_j \text{ is transit;} \\[4mm] \varphi - \dfrac{\langle b\varphi \rangle_Q}{\langle b \rangle_Q} & \text{if } Q_j \text{ is terminal} \end{cases}$$

$(j = 1, \dots, 2^n)$.

We apply to these operators exactly the same proof which led us to (12.13). It is applied verbatim. And we get (12.15). The convergence of $\Lambda^* f + \sum_Q \delta_Q^* f$ to f in $L^2(\mu)$ follows immediately from the fact that partial sums are pointwisely estimated by the dyadic maximal function of f. But dyadic maximal operator is known to be bounded for any measure. So we have the summable majorant, and the by Lebesgue Dominant Convergence theorem we get our strong convergence. The assumptions of Lemma 12.7 are verified. Its application gives us the estimate from below in Lemma 12.5:

$$(12.16) \qquad \|f\|^2 \le \frac{A\hat{A}^2}{\eta^2} \left(\|\Lambda f\|^2 + \sum_{Q \in \text{transit}} \|\Delta_Q f\|^2 \right).$$

Lemma 12.5 is completely proved. $\qquad\qquad\qquad\qquad\qquad\qquad\qquad\qquad\qquad$ □

CHAPTER 13

Next Step in Theorem 10.6. Good and Bad Functions

We consider two functions $f, g \in L^2(\mu)$. Fixing $\omega = (\omega_1, \omega_2)$ we fix two dyadic lattices $\mathcal{D}_1$, $\mathcal{D}_2$ satisfying (7.19). Given $\eta > 0$ and $B < \infty$ from Theorem 12.1 we defined a $\mu, b, D = AB, \eta$ decomposition of f, g, $f = \Lambda f + \sum_{Q \in \mathcal{D}_1^{tr}} \Delta_Q f$, $g = \Lambda g + \sum_{R \in \mathcal{D}_2^{tr}} \Delta_R g$.

We recall the notion of good (bad) cube, good (bad) function from Section 10.2; but we change slightly the definition of goodness.

13.1. Good functions and bad functions again

We have two lattices $\mathcal{D}_1$, $\mathcal{D}_2$ satisfying (7.19). First we define what is a **bad cube** and a **good cube**.

Good and bad cubes Fix a small number $\delta > 0$. Set

(13.1) $$r \text{ integer} : 2^{-r} \leq \delta^S < 2^{-r+1},$$

where S is a large number to be chosen (for now think that it is 2).

By skR we denote $\cup_{i=1}^{2^n} \partial R_i$, where R_i are dyadic children of R.

Let ε, m be parameters from (8.7). We fixed forever

$$\alpha := \frac{\varepsilon}{2\varepsilon + 2m}.$$

A cube $Q \in \mathcal{D}_1$ is called *bad* (actually δ-bad) if there exists a cube $R \in \mathcal{D}_2$ such that 1) $\ell(R) \geq 2^r \ell(Q)$, and 2) $\text{dist}(Q, skR) < \ell(Q)^\alpha \ell(R)^{1-\alpha}$. The same definition applies for $R \in \mathcal{D}_2$.

Definition. Naturally, we can say that $\varphi = \sum_{Q \in \mathcal{D}_1^{tr}} \Delta_Q \varphi$ is bad if only bad Q's participate in this decomposition. The same applies for $\psi = \sum_{R \in \mathcal{D}_2^{tr}} \Delta_R \psi$. In particular, given $\omega = (\omega_1, \omega_2) \in \Omega^2$ and μ, b, D, η decomposition of $f = \Lambda f + \sum_{Q \in \mathcal{D}_1^{tr}} \Delta_Q f$, we have the decomposition to good and bad parts:

$$f = f_{good} + f_{bad}, \text{ where } f_{good} = \Lambda f + \sum_{Q \in \mathcal{D}_1^{tr}, Q \, good} \Delta_Q f.$$

The same applies for $g = \Lambda g + \sum_{Q \in \mathcal{D}_1^{tr}} \Delta_Q g$.

Notice that we have Theorem 10.1. In our case the property of being a good cube is less restrictive than in Section 10.2 (we do not require well-intersectedness to be good), so the probability of being bad is even smaller than in Theorem 10.1. Hence, one can choose $S = S(\alpha)$ in such a way that for any fixed $Q \in \mathcal{D}_1$,

(13.2) $$\mathbb{P}_{\omega_2}\{Q \text{ is bad}\} \leq \delta^2.$$

We are now ready to prove

THEOREM 13.1. *Let function b satisfy (7.5). We consider μ, b, D, η decomposition of f, and take a bad part of it for every $\omega = (\omega_1, \omega_2) \in \Omega^2$. One can choose $S = S(\alpha)$ in such a way that*

$$(13.3) \qquad \mathbb{E}(\|f_{bad}\|) \leq C(\hat{A}, \eta)\delta\|f\| \, .$$

The proof depends only on the property (13.2) and not on a particular definition of badness and goodness.

Proof. By Lemma 12.5 (its left inequality),

$$\mathbb{E}(\|f_{bad}\|) \leq C(\hat{A}, \eta)\mathbb{E}\Big(\sum_{Q \in \mathcal{D}_1^{tr}, \, Q \, bad} \|\Delta_Q f\|^2 \Big)^{1/2} \, .$$

Then

$$\mathbb{E}(\|f_{bad}\|) \leq C(\hat{A}, \eta)\Big(\mathbb{E} \sum_{Q \in \mathcal{D}_1^{tr}, \, Q \, bad} \|\Delta_Q f\|^2 \Big)^{1/2} \, .$$

Let Q be a fixed cube in $\mathcal{D}_1$; then, using (13.2), we conclude:

$$\mathbb{E}_{\omega_2}\|\Delta_Q f\|^2 = \mathbb{P}_{\omega_2}\{Q \text{ is bad}\}\|\Delta_Q f\|^2 \leq \delta^2\|\Delta_Q f\|^2 \, .$$

Therefore, we can continue as follows:

$$\mathbb{E}(\|f_{bad}\|) \leq C(\hat{A}, \eta)\delta\Big(\sum_{Q \in \mathcal{D}_1^{tr}, \, Q \, bad} \|\Delta_Q f\|^2 \Big)^{1/2} \leq C(\hat{A}, \eta)\delta\|f\| \, .$$

The last inequality uses Lemma 12.5 again (its right inequality).   □

13.2. Reduction to estimates on good functions

We consider $\omega = (\omega_1, \omega_2) \in \Omega^2$, μ, b, D, η decomposition of two functions $\varphi, \psi \in L^2(\mu)$ with respect to $\mathcal{D}_1, \mathcal{D}_2$ respectively, and consider the good and bad parts of these decompositions.

Using $\varphi = \varphi_{good} + \varphi_{bad}, \psi = \psi_{good} + \psi_{bad}$ we can write

$$\langle T\varphi, \psi \rangle = \langle T\varphi_{good}, \psi_{good} \rangle + \langle T\varphi_{bad}, \psi_{good} \rangle + \langle T\varphi, \psi_{bad} \rangle \, .$$

Call the sum of the last two terms $R_1(\omega, \varphi, \psi)$.

THEOREM 13.2. *Let T be any operator with bounded kernel. Then*

$$\mathbb{E}R_1(\omega, \varphi, \psi) \leq C(\hat{A}, \eta)\|T\|\delta\|\varphi\|\|\psi\| \, .$$

Proof. The proof is obvious. Just use Theorem 13.1.   □

Having in mind that we have to prove (12.1), we are left to prove that if φ, ψ are two δ-good functions, then

(13.4)
$$\varphi, \psi \text{ are two good functions} \Rightarrow |\langle T\varphi, \psi \rangle| \leq A \, C \, \delta^{-A}\|\varphi\|\|\psi\| + \|T\|R(\omega, \varphi, \psi) \, ,$$

where C is a finite constant depending only on the parameters of Theorem 12.1, and where the remainder satisfies (12.6).

13.3. Splitting $\langle T\varphi_{good}, \psi_{good}\rangle$ to three sums

Getting rid of Λ. From now on we consider $\omega = (\omega_1, \omega_2) \in \Omega^2$, μ, b, D, η decomposition of two functions $\varphi, \psi \in L^2(\mu)$ with respect to $\mathcal{D}_1, \mathcal{D}_2$ respectively, our φ, ψ are δ-good functions (if otherwise not stated).

Note first of all, that it is enough to prove the desired inequality for functions φ and ψ such that $\Lambda\varphi = \Lambda\psi = 0$.

Indeed, for any $\varphi \in L^2(\mu)$, we have

$$\|T\Lambda\varphi\|_{L^2(\mu)} = \frac{|\langle\varphi\rangle_{Q^0}|}{|\langle b\rangle_{Q^0}|}\|Tb\|_{L^2(\mu)} \leq \frac{1}{\eta}|\langle\varphi\rangle_{Q^0}| \cdot \|Tb\|_{L^\infty(\mu)}\sqrt{\mu(Q^0)}$$

$$\leq \frac{C_6}{\eta}|\langle\varphi\rangle_{Q^0}|\sqrt{\mu(Q^0)} \leq \frac{C_6}{\eta}\|\varphi\|_{L^2(\mu)},$$

where we used the constant C_6 from (12.2). Taking into account that $\langle\varphi, T\psi\rangle = -\langle T\varphi, \psi\rangle$ for all $\varphi, \psi \in L^2(\mu)$, we get

$$\langle\varphi, T\psi\rangle = -\langle T\Lambda\varphi, \psi\rangle + \langle\varphi - \Lambda\varphi, T\Lambda\psi\rangle + \langle\varphi - \Lambda\varphi, T(\psi - \Lambda\psi)\rangle.$$

The first two terms do not exceed $\frac{C_6}{\eta}\|\varphi\|_{L^2(\mu)}\|\psi\|_{L^2(\mu)}$ and $4B\|\varphi\|_{L^2(\mu)}\|\psi\|_{L^2(\mu)}$, correspondingly (because $\|\varphi - \Lambda\varphi\|_{L^2(\mu)} \leq 2\|\varphi\|_{L^2(\mu)}$). Meanwhile, the functions $\varphi' = \varphi - \Lambda\varphi$ and $\psi' = \psi - \Lambda\psi$ clearly satisfy the condition $\Lambda\varphi = \Lambda\psi = 0$ and their $L^2(\mu)$-norms are bounded by $2\|\varphi\|_{L^2(\mu)}$ and $2\|\varphi\|_{L^2(\mu)}$, respectively. So we reduced the estimate (12.5) to the case of no $\Lambda f, \Lambda g$. Of course, now

$$\langle Tf, g\rangle = \sum_{Q\,good, R\,good\,\ell(Q)\leq\ell(R)} \langle\Delta_Q f, \Delta_R g\rangle$$

$$+ \sum_{Q\,good, R\,good\,\ell(Q)\geq\ell(R)} \langle\Delta_Q f, \Delta_R g\rangle.$$

We would like to write

$$\langle\varphi, T\psi\rangle = \sum_{Q\in\mathcal{D}_1^{tr}, R\in\mathcal{D}_2^{tr}} \langle\Delta_Q\varphi, T\Delta_R\psi\rangle.$$

The question arises as to why this series converges in any reasonable sense. But let us observe that, since the operator T has bounded kernel, the operator T is bounded in $L^2(\mu)$ and therefore we can restrict ourselves to the good functions φ and ψ that have only finitely many non-zero terms in their decompositions (clearly, if φ is good, then any partial sum of the series $\Lambda\varphi + \sum_{Q\in\mathcal{D}_1}\Delta_Q\varphi$ is good as well). This not only removes any questions about the convergence, but also allows us to rearrange and to group the terms in the sum in any way we want.

Due to this observation and due to the (anti)symmetry, it is enough to estimate the sum over $Q \in \mathcal{D}_1^{tr}$ and $R \in \mathcal{D}_2^{tr}$, for which $\ell(Q) \leq \ell(R)$.

Notation. For the sake of notational simplicity, everywhere below instead of

$$\sum_{Q\in\mathcal{D}_1^{tr}, Q\text{ is good}, R\in\mathcal{D}_2^{tr}, \ell(Q)\leq\ell(R), \text{other conditions}},$$

we will write

$$\sum_{Q,R:\text{ other conditions}}.$$

Also we will always reduce $\sum_{Q \in \mathcal{D}_1^{tr}:\, Q \text{ is good, other conditions}}$ to $\sum_{Q:\text{ other conditions}}$ and $\sum_{R \in \mathcal{D}_2^{tr}:\text{ other conditions}}$ to $\sum_{R:\text{ other conditions}}$.

Note, that (unless otherwise specified) we will always think that the summation over Q goes only over *good* cubes $Q \in \mathcal{D}_1^{tr}$, while the summation over R goes over *all* $R \in \mathcal{D}_2^{tr}$.

Of course, formally it doesn't matter, because, since the functions φ and ψ are good, it is merely a business of adding or omitting several zeros. But it will allow us (and the reader) to see clearly where and what property is used. As the reader might have already guessed, for the sum over pairs Q, R with $\ell(Q) > \ell(R)$, this point of view should be changed to the opposite.

The definition of δ-goodness involved a large integer $r = r(\delta)$ (see Section 13.1). Use it to write (we recall for the last time that according to our notation $\ell(Q) \leq \ell(R)$ everywhere below):

$$\sum_{Q,R} \langle \Delta_Q \varphi, T \Delta_R \psi \rangle = \sum_{Q,R:\ell(Q) \geq 2^{-r}\ell(R)} + \sum_{Q,R:\ell(Q) < 2^{-r}\ell(R)}$$

$$= \sum_{\substack{Q,R:\ell(Q) \geq 2^{-r}\ell(R), \\ \text{dist}(Q,R) \leq \ell(R)}} + \left[\sum_{\substack{Q,R:\ell(Q) \geq 2^{-r}\ell(R), \\ \text{dist}(Q,R) > \ell(R)}} + \sum_{\substack{Q,R:\ell(Q) < 2^{-r}\ell(R), \\ Q \cap R = \emptyset}} \right] + \sum_{\substack{Q,R:\ell(Q) < 2^{-r}\ell(R), \\ Q \cap R \neq \emptyset}}$$

$$:= \sigma_1 + \sigma_2 + \sigma_3.$$

Definition We call the first sum *the diagonal sum*, the second one *the long range interaction sum*, and the third one *the short range interaction sum*.

13.4. Three types of estimates of $\int k(x,y) f(x) g(y) \, d\mu(x) \, d\mu(y)$

Recall that the kernel $k(x,y)$ of T satisfies the estimates (12.3), (12.4) and

$$|k(x,y)| \leq \frac{C_{CZ}}{|x-y|^m}.$$

The second inequality implies that

$$|k(x,y) - k(x',y)| \leq \frac{C_{CZ}|x-x'|^\tau}{|x-y|^{m+\tau}}$$

provided that $|x - x'| \leq \frac{1}{2}|x - y|$, with some (fixed) $0 < \varepsilon \leq 1$ and $0 < C_{CZ} < \infty$.

First of all we will sometimes write

$$\int \int k(x,y) f(x) g(y) \, d\mu(x) \, d\mu(y) = \int \int [k(x,y) - k(x_0,y)] f(x) g(y) \, d\mu(x) \, d\mu(y)$$

using the fact that our f, g will be $\Delta_Q \varphi$, $\Delta_R \psi$ and so their integrals are zero. Temporarily call $K(x,y)$ either $k(x,y)$ or $k(x,y) - k(x_0, y)$.

After that we have three logical possibilities to estimate

$$\int \int K(x,y) f(x) g(y) \, d\mu(x) \, d\mu(y).$$

1) estimate $|K|$ in L^∞, and f, g in L^1 norms; 2) estimate $|K|$ in $L^\infty L^1$ norm, and f in L^1 norm, g in L^∞ norm (or maybe, do this symmetrically); 3) estimate $|K|$ in L^1 norm, and f, g in L^∞ norms.

The third method is widely used for Calderón–Zygmund estimates on homogeneous spaces (say with respect to Lebesgue measure), but it is very dangerous to use it for nonhomogeneous measure. Here is the reason. After f, g are estimated in L^∞ norms, one needs to continue these estimates to have L^2 norms. There is nothing strange in that as usually f, g are almost proportional to characteristic functions. But for f living on Q such that $f = c_Q \chi_Q$ (c_Q is a constant),

$$\|f\|_{L^\infty(\mu)} \leq \frac{1}{(\mu(Q))^{1/2}} \|f\|_{L^2(\mu)} \, .$$

The same reasoning applies for g on R. Then

$$\left| \int \int K(x,y)f(x)g(y)\,d\mu(x)\,d\mu(y) \right| \leq \frac{1}{(\mu(Q))^{1/2}(\mu(R))^{1/2}} \|f\|_{L^2(\mu)} \|g\|_{L^2(\mu)} \, .$$

And nonhomogeneous measure has no estimate from below. Having two uncontrollable almost zeroes in the denominator is a very bad idea. We will never use the estimate of type 3).

On the other hand, estimate of the type 2) is much less dangerous (although requires the care as well). This is because in this case one applies

$$\|f\|_{L^1(\mu)} \leq (\mu(Q))^{1/2} \|f\|_{L^2(\mu)} \, , \|g\|_{L^\infty(\mu)} \leq \frac{1}{(\mu(Q))^{1/2}} \|g\|_{L^2(\mu)} \, ,$$

and gets

$$\left| \int \int K(x,y)f(x)g(y)\,d\mu(x)\,d\mu(y) \right| \leq \frac{(\mu(Q))^{1/2}}{(\mu(R))^{1/2}} \|f\|_{L^2(\mu)} \|g\|_{L^2(\mu)} \, .$$

If we choose to use estimate of the type 2) only for pairs Q, R such that $Q \subset R$ we are in good shape. This is what we will be doing estimating σ_3.

Plan. We use estimate of the type 1) for long range estimations in Section 13.5. This is relatively easy. Then we use the estimates of type 1) and 2) for our short range estimations in Section 13.5. This is the most important part. Exactly here (as usual in this type of question) the so-called paraproducts appear. This indicates that the sums in the short range interaction part cannot be estimated by putting the absolute values on each of their terms. One should find large blocks, where the cancellation appears.

The third part (see Chapter 14) deals with the diagonal part of the operator. By all means it is supposed to be an easy part of the job, but this is not so here.

Remarks about the diagonal sum. It is quite difficult for three reasons: a) measure μ does not have any estimate from above (it does not have any estimate from below either, but this goes without saying as we work with the nonhomogeneous situation), b) function b is non-accretive, c) n can be greater than 2. If measure μ at least would have the following estimate from above (see also the discussion around (10.9)):

$$(13.5) \qquad\qquad \mu(B(x,r)) \leq C\,r^m \, ,$$

and b would be accretive, then the proofs in Chapter 14 could have been made considerably simpler. One can also have simpler proofs with our bad non-accretive b and with general measures (without (13.5)), but only if one works with $n = 2$.

We called measures with (13.5) measures of order m, and worked with them in [**43**], [**44**], [**45**]. In [**46**] $m = 1$, but the measure under consideration does not satisfy (13.5) uniformly. It is however of linear growth, meaning that for μ almost every x,

$$\limsup_{r \to 0} \frac{\mu(B(x,r))}{r} < \infty .$$

Also in [**46**] the dimension n of the ambient space is equal to 2. This case is slightly easier than the case of $n > 2$. Here all bad things meet together, and one should deal with all of them simultaneously.

13.5. Estimate of long range interaction sum σ_2

Recall that the sum σ_2 is taken over pairs Q, R such that $Q \cap R = \emptyset$.

If $\ell(Q) \geq 2^{-r}\ell(R)$, then the cubes not only do not intersect, but they are also well-separated: $\mathrm{dist}(Q, R) \geq \ell(R)$. We would like to extend this property onto the case $\ell(Q) < 2^{-r}\ell(R)$. Though we cannot achieve exactly the same separation by the length of the larger cube, we can get as close to it as we want. Namely, for any $\alpha > 0$ and for any $Q \in \mathcal{D}_1$, the probability

$$P_{\omega_2}\{\mathcal{D}_2 : \exists R \in \mathcal{D}_2 : \ell(R) > 2^m\ell(Q), R \cap Q = \emptyset \ \text{and} \ \ \mathrm{dist}(Q, R) < \ell(Q)^\alpha\ell(R)^{1-\alpha}\}$$

allows an estimate that does not depend on Q and tends to 0 as $m \to \infty$.

We shall need this result for $\alpha = \frac{\varepsilon}{2(m+\tau)}$. This is proved in Theorem 10.1. Let us observe that if φ is good, then for *every* pair Q, R participating in σ_2, we have $\mathrm{dist}(Q, R) \geq \ell(Q)^\alpha\ell(R)^{1-\alpha}$.

Define the *long distance* $D(Q, R)$ between the cubes Q and R by

$$D(Q, R) = \ell(Q) + \ell(R) + \mathrm{dist}(Q, R).$$

Notation. The letter A stands for constants depending only on the dimension. By the letter C we will temporarily denote the constants that depend only on Calderón–Zygmund parameters of the operator T and a parameter D of our μ, b, D, η decomposition of φ, ψ. We recall that we chose $D = A B$, where B is from (12.3) of the result we are striving to prove: Theorem 12.1.

LEMMA 13.3. *Suppose that Q and R are two cubes on the complex plane $\mathbb{C}$, such that $\ell(Q) \leq \ell(R)$. Let $\varphi_Q, \psi_R \in L^2(\mu)$. Assume that φ_Q vanishes outside Q, ψ_R vanishes outside R; $\int_{\mathbb{C}} \varphi_Q = 0$ and, at last, $\mathrm{dist}(Q, \mathrm{supp}\, \psi_R) \geq \ell(Q)^\alpha\ell(R)^{1-\alpha}$. Then*

$$|\langle \varphi_Q, T\psi_R \rangle| \leq A\, C\, \frac{\ell(Q)^{\frac{\tau}{2}}\ell(R)^{\frac{\tau}{2}}}{D(Q, R)^{m+\tau}}\sqrt{\mu(Q)}\sqrt{\mu(R)}\|\varphi_Q\|_{L^2(\mu)}\|\psi_R\|_{L^2(\mu)}.$$

Remark. Note that we require only that the support of the function ψ lies far from Q; the cubes Q and R themselves may intersect! We will really have such a situation when estimating σ_3.

Proof. Let x_Q be the center of the cube Q. Note that for all $x \in Q$, $y \in \mathrm{supp}\, \psi_R$, we have

$$|x_Q - y| \geq \frac{\ell(Q)}{2} + \mathrm{dist}(Q, \mathrm{supp}\, \psi_R) \geq \frac{\ell(Q)}{2} + 2^{r(1-\alpha)}\ell(Q) \geq A\,\ell(Q) \geq 2|x - x_Q|.$$

Therefore,

$$
\begin{aligned}
|\langle \varphi_Q, T\psi_R \rangle| &= \left| \iint k(x,y)\varphi_Q(x)\psi_R(y)\, d\mu(x)\, d\mu(y) \right| \\
&= \left| \iint [k(x,y) - k(x_Q,y)]\varphi_Q(x)\psi_R(y)\, d\mu(x)\, d\mu(y) \right| \\
&\leq A \frac{\ell(Q)^\tau}{\operatorname{dist}(Q, \operatorname{supp}\psi_R)^{m+\tau}} \|\varphi_Q\|_{L^1(\mu)} \|\psi_R\|_{L^1(\mu)}.
\end{aligned}
$$

There are two possible cases.

Case 1: $\operatorname{dist}(Q, \operatorname{supp}\psi_R) \geq \ell(R)$. Then

$$
D(Q,R) = \ell(Q) + \ell(R) + \operatorname{dist}(Q,R) \leq 3\operatorname{dist}(Q, \operatorname{supp}\psi_R)
$$

and therefore

$$
\frac{\ell(Q)^\tau}{\operatorname{dist}(Q, \operatorname{supp}\psi_R)^{m+\tau}} \leq C \frac{\ell(Q)^\tau}{D(Q,R)^{m+\tau}} \leq C \frac{\ell(Q)^{\frac{\tau}{2}}\ell(R)^{\frac{\tau}{2}}}{D(Q,R)^{m+\tau}}.
$$

Case 2: $\ell(Q)^\alpha \ell(R)^{1-\alpha} \leq \operatorname{dist}(Q, \operatorname{supp}\psi_R) \leq \ell(R)$. Then $D(Q,R) \leq 3\ell(R)$ and we get

$$
\frac{\ell(Q)^\tau}{\operatorname{dist}(Q, \operatorname{supp}\psi_R)^{m+\tau}} \leq \frac{\ell(Q)^\tau}{[\ell(Q)^\alpha \ell(R)^{1-\alpha}]^{m+\tau}} = \frac{\ell(Q)^{\frac{\tau}{2}}\ell(R)^{\frac{\tau}{2}}}{\ell(R)^{m+\tau}} \leq C \frac{\ell(Q)^{\frac{\tau}{2}}\ell(R)^{\frac{\tau}{2}}}{D(Q,R)^{m+\tau}}.
$$

Now, to finish the proof of the lemma, it remains only to note that

$$
\|\varphi_Q\|_{L^1(\mu)} \leq \sqrt{\mu(Q)}\|\varphi_Q\|_{L^2(\mu)} \qquad \text{and} \qquad \|\psi_R\|_{L^1(\mu)} \leq \sqrt{\mu(R)}\|\psi_R\|_{L^2(\mu)}.
$$

Applying this lemma to $\varphi_Q = \Delta_Q\varphi$ and $\psi_R = \Delta_R\psi$, we obtain

$$
(13.6) \qquad |\sigma_2| \leq AC \sum_{Q,R} \frac{\ell(Q)^{\frac{\tau}{2}}\ell(R)^{\frac{\tau}{2}}}{D(Q,R)^{m+\tau}} \sqrt{\mu(Q)}\sqrt{\mu(R)} \|\Delta_Q\varphi\|_{L^2(\mu)} \|\Delta_R\psi\|_{L^2(\mu)}.
$$

$\square$

Notation. We are going to show that the matrix $T_{Q,R}$ defined by

$$
T_{Q,R} := \frac{\ell(Q)^{\frac{\tau}{2}}\ell(R)^{\frac{\tau}{2}}}{D(Q,R)^{m+\tau}} \sqrt{\mu(Q)}\sqrt{\mu(R)} \qquad (Q \in \mathcal{D}_1^{tr},\, R \in \mathcal{D}_2^{tr},\, \ell(Q)\ell(R))
$$

generates a bounded operator in l^2.

LEMMA 13.4. *For any two "sequences" $\{a_Q\}_{Q\in\mathcal{D}_1^{tr}}$ and $\{b_R\}_{R\in\mathcal{D}_2^{tr}}$ of nonnegative numbers, one has*

$$
\sum_{Q,R} T_{Q,R} a_Q b_R \leq AC \left[\sum_Q a_Q^2\right]^{\frac{1}{2}} \left[\sum_R b_R^2\right]^{\frac{1}{2}}.
$$

Remark. Note that $T_{Q,R}$ are defined for all Q,R with $\ell(Q) \leq \ell(R)$ and that the condition $\operatorname{dist}(Q,R) \geq \ell(Q)^\alpha \ell(R)^{1-\alpha}$ (or even the condition $Q \cap R = \emptyset$) no longer appears in the summation!

Proof. Let us "slice" the matrix $T_{Q,R}$ according to the ratio $\frac{\ell(Q)}{\ell(R)}$. Namely, let

$$
T_{Q,R}^{(k)} = \begin{cases} T_{Q,R}, & \text{if } \ell(Q) = 2^{-k}\ell(R); \\ 0, & \text{otherwise}, \end{cases}
$$

$(k = 0, 1, 2, \dots)$. To prove the lemma, it is enough to show that for every $k \geq 0$,

$$\sum_{Q,R} T^{(k)}_{Q,R} a_Q b_R \leq A\,C\,2^{-\frac{\tau}{2}k} \Big[\sum_Q a_Q^2\Big]^{\frac{1}{2}} \Big[\sum_R b_R^2\Big]^{\frac{1}{2}}.$$

The matrix $\{T^{(k)}_{Q,R}\}$ has a "block" structure: the variables b_R corresponding to the cubes $R \in \mathcal{D}^{tr}_2$, for which $\ell(R) = 2^j$, can interact only with variables a_Q corresponding to the cubes $Q \in \mathcal{D}^{tr}_1$, for which $\ell(Q) = 2^{j-k}$. Thus, to get the desired inequality, it is enough to estimate each block separately, i.e., to demonstrate that

$$\sum_{Q,R\,:\,\ell(Q)=2^{j-k},\ell(R)=2^j} T^{(k)}_{Q,R} a_Q b_R \leq A\,C \Big[\sum_{Q\,:\,\ell(Q)=2^{j-k}} a_Q^2\Big]^{\frac{1}{2}} \Big[\sum_{R\,:\,\ell(R)=2^j} b_R^2\Big]^{\frac{1}{2}}.$$

Let us introduce the functions

$$F := \sum_{Q\,:\,\ell(Q)=2^{j-k}} \frac{a_Q}{\sqrt{\mu(Q)}}\chi_Q \quad \text{and} \quad G := \sum_{R\,:\,\ell(R)=2^j} \frac{b_R}{\sqrt{\mu(R)}}\chi_R.$$

Note that the cubes of a given size in one dyadic lattice do not intersect, and therefore at each point $x \in \mathbb{C}$, at most one term in the sum can be non-zero. Also observe that

$$\|F\|_{L^2(\mu)} = \Big[\sum_{Q\,:\,\ell(Q)=2^{j-k}} a_Q^2\Big]^{\frac{1}{2}} \quad \text{and} \quad \|G\|_{L^2(\mu)} = \Big[\sum_{R\,:\,\ell(R)=2^j} b_R^2\Big]^{\frac{1}{2}}.$$

Then the estimate we need can be rewritten as

$$\iint K_{j,k}(x,y)F(x)G(y)\,d\mu(x)\,d\mu(y) \leq A\,C\,\|F\|_{L^2(\mu)}\|G\|_{L^2(\mu)},$$

where

$$K_{j,k}(x,y) = \sum_{Q,R\,:\,\ell(Q)=2^{j-k},\ell(R)=2^j} \frac{\ell(Q)^{\frac{\tau}{2}}\ell(R)^{\frac{\tau}{2}}}{D(Q,R)^{m+\tau}}\chi_Q(x)\chi_R(y).$$

Again, for every pair of points $x, y \in \mathbb{C}$, only one term in the sum can be nonzero. Since $|x-y| + \ell(R) \leq 3D(Q,R)$ for any $x \in Q$, $y \in R$, we obtain

$$K_{j,k}(x,y) = A\,C\,2^{-\frac{\tau}{2}k}\frac{\ell(R)^\tau}{D(Q,R)^{m+\tau}}$$

$$\leq A\,C\,2^{-\frac{\tau}{2}k}\frac{2^{j\tau}}{[2^j + |x-y|]^{m+\tau}} =: A\,C\,2^{-\frac{\tau}{2}k}k_j(x,y).$$

So, it is enough to check that

$$\iint k_j(x,y)F(x)G(y)\,d\mu(x)\,d\mu(y) \leq A\,C\,\|F\|_{L^2(\mu)}\|G\|_{L^2(\mu)}.$$

According to the Schur test, it would suffice to prove that for every $y \in \mathbb{R}^n$, one has the estimate $\int_{\mathbb{C}} k_j(x,y)\,d\mu(x) \leq A\,C$ and vice versa (i.e., for every $x \in \mathbb{C}$, one has $\int_{\mathbb{C}} k_j(x,y)\,d\mu(y) \leq a\,C$). Then the norm of the integral operator with kernel k_j in $L^2(\mu)$ would be bounded by the same constant $A\,C$, and the proof of Lemma 13.4 would be over.

If we assumed a priori that the supremum of radii of all $D = A\,B$ non-Ahlfors balls centered at y were less than 2^{j+1}, then the needed estimate would be next to trivial: we could write

$$\int_{\mathbb{C}} k_j(x,y)\, d\mu(x) = \int_{B(y,2^{j+1})} k_j(x,y)\, d\mu(x) + \int_{\mathbb{R}^n \setminus B(y,2^{j+1})} k_j(x,y)\, d\mu(x)$$

$$\leq D\, 2^{-jm} \mu(B(y,2^{j+1})) + \int_{\mathbb{R}^n \setminus B(y,2^{j+1})} \frac{2^{j\tau}}{|x-y|^{m+\tau}}\, d\mu(x)$$

$$\leq D\Big(A + A \int_{2^j}^{+\infty} \frac{2^{j\tau} t^{m-1}}{t^{m+\tau}}\, dt\Big) = A\, C$$

(we use our notational convention that A depends on the dimension and that C depends on parameters of Theorem 12.1).

The problem is that we cannot guarantee that the supremum of radii of all $D = A\, B$ non-Ahlfors balls centered at y be less than 2^{j+1} for every $y \in \mathbb{R}^n$. Our measure may not have (10.9). So, generally speaking, we are unable to show that the integral operator with kernel $k_j(x,y)$ acts in $L^2(\mu)$. But we *do not need* that much! We only need to check that the corresponding bilinear form is bounded on two *given* functions F and G. So, we are not interested in the points $y \in \mathbb{R}^n$ for which $G(y) = 0$ (or in the points $x \in \mathbb{C}$, for which $F(x) = 0$). But, by definition, G can be non-zero only on transit cubes in $\mathcal{D}_2$. Now let us notice that if $R \in \mathcal{D}_2^{tr}$, then the supremum of radii of all $D = A\, B$ non-Ahlfors balls centered at $y \in R$ is bounded by $A\ell(R)$ for every $y \in R$. Indeed, this is just Lemma 12.4.

The same reasoning shows that if $Q \in \mathcal{D}_1^{tr}$, then the supremum of radii of all $D = A\, B$ non-Ahlfors balls centered at $x \in Q$ is bounded by $2^{j-k+1} \leq 2^{j+1}$ whenever $F(x) \neq 0$, and we are done with Lemma 13.4 and with $|\sigma_2|$. $\qquad\square$

Now, we hope, the reader will agree that the decision to declare the cubes contained in H terminal was a good one. As a result, the fact that the measure μ is not Ahlfors did not put us in any real trouble–we just hardly have a chance to notice this fact at all. Also, it is clear why the cubes with small average of b have been declared terminal: this allowed us to treat b like an accretive function all the time.

But it still remains unexplained why we were so eager to have extra conditions (12.3), (12.4) on our Calderón–Zygmund kernel (and to have k satisfying these extra conditions one had to suppress the Calderón–Zygmund kernel on every terminal cube during the third reduction in Chapter 9). The answer is found in the next two sections.

13.6. Short range interaction sum σ_3. Nonhomogeneous paraproducts

Recall that the sum σ_3 is taken over the pairs Q, R, for which $\ell(Q) < 2^{-r}\ell(R)$ and $Q \cap R \neq \emptyset$. We would like to improve this condition to the demand that Q lie "deep inside" one of the four subcubes R_j ($j = 1,2,3,4$). Recall also that we define the *skeleton* sk$R$ of the cube R by

$$\text{sk}R := \bigcup_{j=1}^{2^n} \partial R_j.$$

We have declared a cube $Q \in \mathcal{D}_1$ bad if there exists a cube $R \in \mathcal{D}_2$ such that $\ell(R) > 2^r \ell(Q)$ and $\text{dist}(Q, \text{sk}R) \leq \ell(Q)^\alpha \ell(R)^{1-\alpha}$. See Sections 10.2, 13.1.

Now, for every good cube $Q \in \mathcal{D}_1$, the conditions $\ell(Q) < 2^{-r}\ell(R)$ and $Q \cap R \neq \emptyset$ together imply that Q lies inside one of the 2^n children R_j of R. We will denote this subcube by R_Q. The sum σ_3 can now be split into

$$\sigma_3^{term} := \sum_{\substack{Q,R\,:\,Q\subset R,\,\ell(Q)<2^{-r}\ell(R),\\ R_Q \text{ is terminal}}} \langle \Delta_Q \varphi, T\Delta_R \psi \rangle$$

and

$$\sigma_3^{tr} := \sum_{\substack{Q,R\,:\,Q\subset R,\,\ell(Q)<2^{-r}\ell(R),\\ R_Q \text{ is transit}}} \langle \Delta_Q \varphi, T\Delta_R \psi \rangle.$$

Estimation of σ_3^{term}. First of all, write (recall that R_j denote the children of R):

$$\sigma_3^{term} = \sum_{j=1}^{2^n} \sum_{\substack{Q,R\,:\,\ell(Q)<2^{-r}\ell(R),\\ Q\subset R_j\in\mathcal{D}_2^{term}}} \langle \Delta_Q \varphi, T\Delta_R \psi \rangle.$$

Clearly, it is enough to estimate the inner sum for every fixed $j = 1,\ldots,2^n$. Let us do this for $j = 1$. We have

$$\sum_{\substack{Q,R\,:\,\ell(Q)<2^{-r}\ell(R),\\ Q\subset R_1\in\mathcal{D}_2^{term}}} \langle \Delta_Q \varphi, T\Delta_R \psi \rangle = \sum_{R\,:\,R_1\in\mathcal{D}_2^{term}} \sum_{\substack{Q\,:\,\ell(Q)<2^{-r}\ell(R),\\ Q\subset R_1}} \langle \Delta_Q \varphi, T\Delta_R \psi \rangle.$$

Recall that the kernel k of our operator T satisfies rough estimates (12.3), (12.4) above. Roughly speaking, our main idea here is the following. If $R_1 \in \mathcal{D}_2^{term}$, then for all $x \in R_1$, one has by Lemma 12.3

$$R(x) \geq \text{dist}(x, \partial R_1), \text{ or } d(x) \geq \text{dist}(x, \partial R_1).$$

For the points x that lie in the "central part" of R_1, the right hand side is at least $\frac{\ell(R)^m}{2}$. Assume that it is so for *every* point $x \in R_1$. By the way, it is so for *every* point $x \in R_1$ if R_1 is a terminal cube of the sort described in (12.9), in other words, if it has been announced the terminal because $2R_1 \subset H$ (see Lemma 12.3 and the explanations around it). Then (see (12.3), (12.4))

$$(13.7) \quad k(x,y) \leq \min\left[\frac{1}{R(x)^m}, \frac{1}{\text{dist}(x,\partial\mathcal{T}_{12})}\right] \leq \frac{A}{\ell(R)^m} \qquad \text{for all } x \in R_1, y \in \mathbb{R}^n.$$

Hence

$$(13.8) \qquad |T\Delta_R \psi(x)| \leq \frac{A\|\Delta_R \psi\|_{L^1(\mu)}}{\ell(R)^m} \qquad \text{for all } x \in R_1,$$

and therefore

$$\|\chi_{R_1} \cdot T\Delta_R \psi\|_{L^2(\mu)} \leq A\|\Delta_R \psi\|_{L^1(\mu)} \frac{\sqrt{\mu(R_1)}}{\ell(R)^m}$$

$$\leq \frac{A\mu(R)}{\ell(R)^m}\|\Delta_R \psi\|_{L^2(\mu)} \leq A\,B\|\Delta_R \psi\|_{L^2(\mu)},$$

because $\|\Delta_R \psi\|_{L^1(\mu)} \leq \sqrt{\mu(R)}\|\Delta_R \psi\|_{L^2(\mu)}$, $\mu(R_1) \leq \mu(R)$, and

$$(13.9) \qquad \mu(R) \leq A\,B\,\ell(R)$$

because R (father of R_1) is transit, if R_1 is terminal, and for the transit cubes we have Lemma 12.4.

Now, recalling Lemma 12.5, and taking into account that $\Delta_Q\varphi \equiv 0$ outside Q, we get

$$\sum_{Q:\, Q \subset R_1} |\langle \Delta_Q\varphi, T\Delta_R\psi \rangle| = \sum_{Q:\, Q \subset R_1} |\langle \Delta_Q\varphi, \chi_{R_1} \cdot T\Delta_R\psi \rangle|$$

$$\leq \frac{A\,\hat{A}^2}{\eta^2} \|\chi_{R_1} \cdot T\Delta_R\psi\|_{L^2(\mu)} \Big[\sum_{Q:\, Q \subset R_1} \|\Delta_Q\varphi\|^2_{L^2(\mu)} \Big]^{\frac{1}{2}}$$

$$\leq \frac{A\,\hat{A}^2}{\eta^2} \delta \|\Delta_R\psi\|_{L^2(\mu)} \Big[\sum_{Q:\, Q \subset R_1} \|\Delta_Q\varphi\|^2_{L^2(\mu)} \Big]^{\frac{1}{2}}.$$

So, we obtain

$$\sum_{R:\, R_1 \in \mathcal{D}_2^{term}} \sum_{Q:\, Q \subset R_1} |\langle \Delta_Q\varphi, T\Delta_R\psi \rangle|$$

$$\leq \frac{A\,\hat{A}^2}{\eta^2} \delta \sum_{R:\, R_1 \in \mathcal{D}_2^{term}} \|\Delta_R\psi\|_{L^2(\mu)} \Big[\sum_{Q:\, Q \subset R_1} \|\Delta_Q\varphi\|^2_{L^2(\mu)} \Big]^{\frac{1}{2}}$$

$$\leq \frac{A\,\hat{A}^2}{\eta^2} \delta \Big[\sum_{R:\, R_1 \in \mathcal{D}_2^{term}} \|\Delta_R\psi\|^2_{L^2(\mu)} \Big]^{\frac{1}{2}} \Big[\sum_{R:\, R_1 \in \mathcal{D}_2^{term}} \sum_{Q:\, Q \subset R_1} \|\Delta_Q\varphi\|^2_{L^2(\mu)} \Big]^{\frac{1}{2}}.$$

But the terminal cubes in $\mathcal{D}_2$ do not intersect! Therefore every $\Delta_Q\varphi$ can appear at most once in the last double sum, and we get the bound

$$\sum_{R:\, R_1 \in \mathcal{D}_2^{term}} \sum_{Q:\, Q \subset R_1} |\langle \Delta_Q\varphi, T\Delta_R\psi \rangle|$$

$$\leq \frac{A\,\hat{A}^2}{\eta^2} \delta \Big[\sum_R \|\Delta_R\psi\|^2_{L^2(\mu)} \Big]^{\frac{1}{2}} \Big[\sum_Q \|\Delta_Q\varphi\|^2_{L^2(\mu)} \Big]^{\frac{1}{2}} \leq \frac{A\,\hat{A}^4}{\eta^4} \delta \|\varphi\|_{L^2(\mu)} \|\psi\|_{L^2(\mu)}.$$

Lemma 12.5 has been used again in the last inequality.

The problem is that we cannot guarantee the estimate $R(x) \geq \frac{\ell(R)}{A}$ for *every* point $x \in R_1$. This can be guaranteed only if terminal cubes are only of one sort, namely, the one described in (12.9). In other words, this can be guaranteed only if we do not have the cubes that are η non-accretive. Here is why we have a following important remark.

Remark on non-accretive terminal cubes. If there are no η non-accretive terminal cubes, for example, if b is an η accretive function, then the estimate of σ_3^{term} would already be achieved. We would not need the following consideration.

But if we have non-accretivity of b, the kernel k can grow near the boundary of terminal cube R_1. Nevertheless, due to our definition of good cubes, we need only to consider the cubes $Q \subset R_1$, for which $\text{dist}(Q, \partial R_1) \geq \ell(Q)^\alpha \ell(R)^{1-\alpha}$. So, if such a cube Q lies close to the boundary of R_1, the size $\ell(Q)$ has to be very small and the corresponding function $\Delta_Q\varphi$ should oscillate very fast. We may hope that this fast oscillation will compensate for the growth of the kernel. To show that it is really the case, we need one more standard technical tool.

Whitney decomposition.

Let S^0 be an arbitrary cube in $\mathbb{R}^n$. Consider the standard dyadic lattice starting with the cube S^0, and denote by $W(S^0)$ the family of all maximal subcubes S in this lattice, for which $\operatorname{dist}(S, \partial S^0) \geq l(S)$. The Whitney decomposition $W(S^0)$ has the following remarkable properties: *1) The cubes $S \in W(S^0)$ are pairwise disjoint and cover the interior of S^0; 2) $\operatorname{dist}(S, \partial S^0) = l(S)$ for every $S \in W(S^0)$; 3) The expanded cubes $\widetilde{S} := 2S$ ($S \in W(S^0)$) still lie "deep inside" S;* namely, $\operatorname{dist}(\widetilde{S}, \partial S^0) = \frac{l(S)}{2} = \frac{l(\widetilde{S})}{4}$, *and every point $x \in \mathbb{R}^n$ belongs to at most A cubes $\widetilde{S}$.* Denote again the center of a cube Q by x_Q. For $S \in W(R_1)$, put

$$\psi_{R,S} := \chi_{\widetilde{S}} \Delta_R \psi \qquad \text{and} \qquad \widetilde{\psi}_{R,S} := \chi_{R \setminus \widetilde{S}} \Delta_R.$$

We have

$$\sum_{\substack{Q:\, \ell(Q) < 2^{-r}\ell(R), \\ Q \subset R_1}} \langle \Delta_Q \varphi, T \Delta_R \psi \rangle = \sum_{S \in W(R_1)} \sum_{\substack{Q:\, \ell(Q) < 2^{-r}\ell(R), \\ Q \subset R_1,\, x_Q \in S}} \langle \Delta_Q \varphi, T \Delta_R \psi \rangle$$

$$= \sum_{S \in W(R_1)} \sum_{\substack{Q:\, \ell(Q) < 2^{-r}\ell(R), \\ Q \subset R_1,\, x_Q \in S}} \langle \Delta_Q \varphi, T \psi_{R,S} \rangle$$

$$+ \sum_{S \in W(R_1)} \sum_{\substack{Q:\, \ell(Q) < 2^{-r}\ell(R), \\ Q \subset R_1,\, x_Q \in S}} \langle \Delta_Q \varphi, T \widetilde{\psi}_{R,S} \rangle.$$

Note now that for every good $Q \subset R_1$ such that $x_Q \in S \in W(R_1)$, one has (recall that r is a large integer)

$$8\ell(Q) \leq 2^{r(1-\alpha)}\ell(Q) \leq 2^{r(1-\alpha)}\ell(Q)^\alpha \ell(R)^{1-\alpha}$$
$$\leq \operatorname{dist}(Q, \partial R_1) \leq \operatorname{dist}(x_Q, \partial R_1) \leq 2l(S),$$

and therefore

$$\operatorname{dist}(Q, \operatorname{supp} \widetilde{\psi}_{R,S}) \geq \operatorname{dist}(Q, \partial \widetilde{S}) \geq \frac{l(S) - \ell(Q)}{2} \geq \frac{l(S)}{4} \geq \ell(Q)^\alpha \ell(R)^{1-\alpha}.$$

Now Lemma 13.3 yields

$$|\langle \Delta_Q \varphi, T \widetilde{\psi}_{R,S} \rangle| \leq A \frac{\ell(Q)^{\frac{\tau}{2}} \ell(R)^{\frac{\tau}{2}}}{D(Q,R)^{m+\tau}} \sqrt{\mu(Q)} \sqrt{\mu(R)} \|\Delta_Q \varphi\|_{L^2(\mu)} \|\widetilde{\psi}_{R,S}\|_{L^2(\mu)}.$$

Taking into account that $\|\widetilde{\psi}_{R,S}\|_{L^2(\mu)} \leq \|\Delta_R \psi\|_{L^2(\mu)}$ and summing over all $R \in \mathcal{D}_2^{tr}$, we arrive at the same sum as in the long term interaction of Chapter 13.5 (actually, we arrive at the part of that sum which *has not been used yet*, but *has already been estimated* there).

So, it remains to find a good upper bound for

$$\sum_{S \in W(R_1)} \sum_{\substack{Q:\, \ell(Q) < 2^{-r}\ell(R), \\ Q \subset R_1,\, x_Q \in S}} \langle \Delta_Q \varphi, T \psi_{R,S} \rangle.$$

Observe once more that Q is necessarily transit here (the reader should not forget that $\Delta_Q \varphi$ was defined *only* for transit Q; for other cubes it does not exist, or rather equals zero). Observe then that the conditions $Q \in \mathcal{D}_1^{tr}$, Q is good, $\ell(Q) < 2^{-r}\ell(R)$,

$Q \subset R_1$, and $x_Q \in S$ together imply $Q \subset \widetilde{S}$ (as we have seen above, they even imply that Q lies deep inside $\widetilde{S}$). So, it is enough to estimate the sum

$$(13.10) \qquad \sum_{S \in W(R_1)} \sum_{Q: Q \subset \widetilde{S}, \, x_Q \in S} |\langle \Delta_Q \varphi, T\psi_{R,S}\rangle|.$$

Note now that for *every* $x \in \widetilde{S}$, we have by (12.3), (12.4)

$$|k(x,y)| \leq \frac{1}{\text{dist}(x, \partial R_1)^m} \leq \frac{A}{\delta(\widetilde{S})^m}.$$

Recall that the "naive" reasoning from (13.8) cannot be used for the whole R_1. But it can be used for $\widetilde{S}$. Repeating our "naive" reasoning from (13.7), (13.8) for the cube $\widetilde{S}$ instead of the whole R_1, we obtain

$$\sum_{Q: Q \subset \widetilde{S}, \, x_Q \in S} |\langle \Delta_Q \varphi, T\psi_{R,S}\rangle| \leq \|\chi_{\widetilde{S}} \cdot T\psi_{R,S}\|_{L^2(\mu)} \Big[\sum_{Q: Q \subset \widetilde{S}, \, x_Q \in S} \|\Delta_Q \varphi\|^2_{L^2(\mu)} \Big]^{\frac{1}{2}}$$

$$\leq \frac{A\hat{A}^2}{\eta^2} \frac{\mu(\widetilde{S})}{\delta(\widetilde{S})} \|\psi_{R,S}\|_{L^2(\mu)} \Big[\sum_{Q: Q \subset \widetilde{S}, \, x_Q \in S} \|\Delta_Q \varphi\|^2_{L^2(\mu)} \Big]^{\frac{1}{2}}.$$

We would like to say again that $\mu(\widetilde{S}) \leq AB\,\ell(\widetilde{S})^m$. We had such a conclusion in (13.9). But there we dealt with a terminal cube R_1, whose father (of course) is transit. Here we have to be slightly more careful as S and $\widetilde{S}$ are neither transit nor terminal. They are not dyadic cubes in the first place!

However, it is enough to estimate the sum (13.10) only in the case when this sum is nonzero. If it is nonzero, then necessarily, there exists a transit cube $Q \in \mathcal{D}_1$ lying inside $\widetilde{S}$. Then necessarily $2Q$ intersect ∂H (see the definition of terminal cubes). Then $2\widetilde{S} \cap \partial H \neq \emptyset$. But ∂H consists of points such that every ball centered at them is AB Ahlfors. So there exists a constant A such that

$$(13.11) \qquad \mu(\widetilde{S}) \leq AB\,\ell(S)^m.$$

So, as before, despite the fact that we cannot use the Ahlfors condition *whenever we want to*, we can use it *whenever we need to*.

Thus, we get (using Lemma 12.5)

$$\sum_{\substack{S \in W(R_1)}} \sum_{\substack{Q: Q \subset \widetilde{S}, \\ x_Q \in S}} |\langle \Delta_Q \varphi, T\psi_{R,S}\rangle|$$

$$\leq \frac{A\hat{A}^2 B}{\eta^2} \sum_{S \in W(R_1)} \|\psi_{R,S}\|_{L^2(\mu)} \Big[\sum_{\substack{Q: Q \subset \widetilde{S}, \\ x_Q \in S}} \|\Delta_Q \varphi\|^2_{L^2(\mu)} \Big]^{\frac{1}{2}}$$

$$\leq \frac{A\hat{A}^2 B}{\eta^2} \Big[\sum_{S \in W(R_1)} \|\psi_{R,S}\|^2_{L^2(\mu)} \Big]^{\frac{1}{2}} \Big[\sum_{S \in W(R_1)} \sum_{Q: Q \subset R_1, \, x_Q \in S} \|\Delta_Q \varphi\|^2_{L^2(\mu)} \Big]^{\frac{1}{2}}$$

(we relaxed the condition $Q \subset \widetilde{S}$ in the last sum to $Q \subset R_1$; this causes no harm now). But

$$\sum_{S \in W(R_1)} \|\psi_{R,S}\|_{L^2(\mu)}^2 = \sum_{S \in W(R_1)} \int_{\widetilde{S}} |\Delta_R \psi|^2 \, d\mu \leq A \int_{\mathbb{C}} |\Delta_R \psi|^2 \, d\mu = A \|\Delta_R \psi\|_{L^2(\mu)}^2$$

(because every point lies in not more than A cubes $\widetilde{S}$).

Meanwhile,

$$\sum_{S \in W(R_1)} \sum_{Q: Q \subset R_1, \, x_Q \in S} \|\Delta_Q \varphi\|_{L^2(\mu)}^2 = \sum_{Q: Q \subset R_1} \|\Delta_Q \varphi\|_{L^2(\mu)}^2.$$

Hence, summing over all $R \in \mathcal{D}_2^{tr}$, for which $R_1 \in \mathcal{D}_2^{term}$, we get

$$\sum_{R: R_1 \in \mathcal{D}_2^{term}} \sum_{S \in W(R_1)} \sum_{\substack{Q: Q \subset \widetilde{S}, \\ x_Q \in S}} |\langle \Delta_Q \varphi, T\psi_{R,S} \rangle|$$

$$\leq \frac{A\hat{A}^2 B}{\eta^2} \sum_{R: R_1 \in \mathcal{D}_2^{term}} \|\Delta_R \psi\|_{L^2(\mu)} \left[\sum_{Q: Q \subset R_1} \|\Delta_Q \varphi\|_{L^2(\mu)}^2 \right]^{\frac{1}{2}}$$

$$\leq \frac{A\hat{A}^2 B}{\eta^2} \left[\sum_{R: R_1 \in \mathcal{D}_2^{term}} \|\Delta_R \psi\|_{L^2(\mu)}^2 \right]^{\frac{1}{2}} \left[\sum_{R: R_1 \in \mathcal{D}_2^{term}} \sum_{Q: Q \subset R_1} \|\Delta_Q \varphi\|_{L^2(\mu)}^2 \right]^{\frac{1}{2}}$$

$$\leq \frac{A\hat{A}^2 B}{\eta^2} \left[\sum_R \|\Delta_R \psi\|_{L^2(\mu)}^2 \right]^{\frac{1}{2}} \left[\sum_Q \|\Delta_Q \varphi\|_{L^2(\mu)} \right]^{\frac{1}{2}}$$

$$\leq \frac{A\hat{A}^4 B}{\eta^4} \|\varphi\|_{L^2(\mu)} \|\psi\|_{L^2(\mu)},$$

finishing the story about σ_3^{term}. The last inequality is based on Lemma 12.5.

Estimation of σ_3^{tr}. Recall that

$$\sigma_3^{tr} = \sum_{\substack{Q,R: Q \subset R, \, \ell(Q) < 2^{-r}\ell(R), \\ R_Q \text{ is transit}}} \langle \Delta_Q \varphi, T\Delta_R \psi \rangle.$$

Split every term in the sum as

$$\langle \Delta_Q \varphi, T\Delta_R \psi \rangle = \langle \Delta_Q \varphi, T(\chi_{R_Q} \Delta_R \psi) \rangle + \langle \Delta_Q \varphi, T(\chi_{R \setminus R_Q} \Delta_R \psi) \rangle.$$

Observe that since Q is good, $Q \subset R$, and $\ell(Q) < 2^{-r}\ell(R)$, we have

$$\text{dist}(Q, \text{supp} \, \chi_{R \setminus R_Q} \Delta_R \psi) \geq \text{dist}(Q, \text{sk}R) \geq \ell(Q)^\alpha \ell(R)^{1-\alpha}.$$

Using Lemma 13.3 and taking into account that the norm $\|\chi_{R \setminus R_Q} \Delta_R \psi\|_{L^2(\mu)}$ does not exceed $\|\Delta_R \psi\|_{L^2(\mu)}$, we conclude that the sum

$$\sum_{\substack{Q,R: Q \subset R, \, \ell(Q) < 2^{-r}\ell(R), \\ R_Q \text{ is transit}}} |\langle \Delta_Q \varphi, T(\chi_{R \setminus R_Q} \Delta_R \psi) \rangle|$$

can be estimated by the sum (13.6) from Chapter 13.5.

Thus, our task is to find a good bound for the sum

$$\sum_{\substack{Q,R:\,Q\subset R,\,\ell(Q)<2^{-r}\ell(R),\\ R_Q \text{ is transit}}} \langle \Delta_Q \varphi, T(\chi_{R_Q} \Delta_R \psi)\rangle.$$

Recalling the definition of $\Delta_R \psi$ and recalling that R_Q is a *transit* cube, we get

$$\chi_{R_Q} \Delta_R \psi = c_{R,Q} \chi_{R_Q} b,$$

where

$$c_{R,Q} = \frac{\langle \psi \rangle_{R_Q}}{\langle b \rangle_{R_Q}} - \frac{\langle \psi \rangle_R}{\langle b \rangle_R}$$

is a *constant*. So, our sum can be rewritten as

$$\sum_{\substack{Q,R:\,Q\subset R,\,\ell(Q)<2^{-r}\ell(R),\\ R_Q \text{ is transit}}} c_{R,Q} \langle \Delta_Q \varphi, T(\chi_{R_Q} b)\rangle.$$

Our next goal will be to extend the function $\chi_{R_Q} b$ to the whole function b in every term (which is exactly the opposite of the idea of the previous section, where, in a similar situation, we tried to "shrink" the function $\Delta_R \psi$ to $\psi_{R,S}$).

Let us observe that

$$\langle \Delta_Q \varphi, T(\chi_{\mathbb{R}^n \setminus R_Q} b)\rangle = \int_{\mathbb{R}^n \setminus R_Q} k(x,y) \Delta_Q \varphi(x) b(y)\, d\mu(x)\, d\mu(y)$$

$$= \int_{\mathbb{R}^n \setminus R_Q} [k(x,y) - k(x_Q, y)] \Delta_Q \varphi(x) b(y)\, d\mu(x)\, d\mu(y).$$

Note again that for every $x \in Q$, $y \in \mathbb{R}^n \setminus R_Q$, we have

$$|x_Q - y| \geq \frac{\ell(Q)}{2} + \text{dist}(Q, \mathbb{R}^n \setminus R_Q) \geq A\ell(Q) \geq 2|x - x_Q|.$$

Therefore

$$|k(x,y) - k(x_Q, y)| \leq \frac{A|x - x_Q|^\tau}{|x_Q - y|^{m+\tau}} \leq \frac{A\ell(Q)^\tau}{|x_Q - y|^{m+\tau}},$$

and

$$|\langle \Delta_Q \varphi, T(\chi_{\mathbb{R}^n \setminus R_Q} b)\rangle| \leq A\ell(Q)^\tau \|\Delta_Q \varphi\|_{L^1(\mu)} \int_{\mathbb{R}^n \setminus R_Q} \frac{|b(y)|\, d\mu(y)}{|x_Q - y|^{m+\tau}}.$$

Now let us consider the sequence of cubes $R^{(j)} \in \mathcal{D}_2$, beginning with $R^{(0)} = R_Q$ and gradually ascending ($R^{(j)} \subset R^{(j+1)}$, $\ell(R^{(j+1)}) = 2\ell(R^{(j)})$) to the starting cube $R^0 = R^{(N)}$ of the lattice $\mathcal{D}_2$. Clearly, all the cubes $R^{(j)}$ are transit.

We have

$$\int_{\mathbb{R}^n \setminus R_Q} \frac{|b(y)|\, d\mu(y)}{|x_Q - y|^{m+\tau}} = \int_{R^0 \setminus R_Q} \frac{|b(y)|\, d\mu(y)}{|x_Q - y|^{m+\tau}} = \sum_{j=1}^N \int_{R^{(j)} \setminus R^{(j-1)}} \frac{|b(y)|\, d\mu(y)}{|x_Q - y|^{m+\tau}}.$$

We call the j-th term here I_j. Note now that, since Q is good and $\ell(Q) < 2^{-r}\ell(R) \leq 2^{-r}\ell(R^{(j)})$ for all $j = 1, \ldots, N$, we have

$$\operatorname{dist}(Q, R^{(j)} \setminus R^{(j-1)}) \geq \operatorname{dist}(Q, sk R^{(j)}) \geq \ell(Q)^\alpha \ell(R^{(j)})^{1-\alpha}.$$

Hence

$$I_j \leq \frac{1}{[\ell(Q)^\alpha \ell(R^{(j)})^{1-\alpha}]^{m+\tau}} \int_{R^{(j)}} |b|\, d\mu.$$

Recalling that $\alpha = \frac{\varepsilon}{2(m+\tau)}$ (see (10.4)), we see that the first factor equals

$$\frac{1}{\ell(Q)^{\frac{\tau}{2}} \ell(R^{(j)})^{m+\frac{\tau}{2}}}.$$

Since $R^{(j)}$ is transit, we have

$$\int_{R^{(j)}} |b|\, d\mu \leq \hat{A}\mu(R^{(j)}) \leq A\,\hat{A}\,B\,\ell(R^{(j)})^m.$$

Thus,

$$I_j \leq \frac{A\,\hat{A}\,B}{\ell(Q)^{\frac{\tau}{2}} \ell(R^{(j)})^{\frac{\tau}{2}}} = 2^{-(j-1)\frac{\varepsilon}{2}} \frac{A\,\hat{A}\,B}{\ell(Q)^{\frac{\tau}{2}} \ell(R)^{\frac{\tau}{2}}}.$$

Summing over $j \geq 1$, we get

$$\int_{\mathbb{R}^n \setminus R_Q} \frac{|b(y)|\, d\mu(y)}{|x_Q - y|^{m+\tau}} = \sum_{j=1}^N I_j \leq \frac{A\,\hat{A}\,B}{1 - 2^{-\frac{\tau}{2}}} \frac{1}{\ell(Q)^{\frac{\tau}{2}} \ell(R)^{\frac{\tau}{2}}}.$$

Now let us note that, since $R_Q \in \mathcal{D}_2^{tr}$, we have

$$\|\Delta_R \psi\|_{L^2(\mu)}^2 \geq \int_{R_Q} |\Delta_R \psi\|^2\, d\mu = |c_{Q,R}|^2 \int_{R_Q} |b|^2\, d\mu$$

$$\geq |c_{Q,R}|^2 |\langle b \rangle|_{R_Q}^2 \mu(R_Q) \geq \eta^2 |c_{Q,R}|^2 \mu(R_Q).$$

So,

$$|c_{Q,R}| \leq \frac{1}{\eta} \frac{\|\Delta_R \psi\|_{L^2(\mu)}}{\sqrt{\mu(R_Q)}}.$$

Combining this estimate with the Cauchy inequality

$$\|\Delta_Q \varphi\|_{L^1(\mu)} \leq \sqrt{\mu(Q)} \|\Delta_Q \varphi\|_{L^2(\mu)},$$

we finally obtain

$$|\langle \Delta_Q \varphi, T(\chi_{\mathbb{R}^n \setminus R_Q} b) \rangle|$$

$$\leq \frac{A\,\hat{A}\,B}{\eta(1 - 2^{-\frac{\tau}{2}})} \left[\frac{\ell(Q)}{\ell(R)}\right]^{\frac{\tau}{2}} \sqrt{\frac{\mu(Q)}{\mu(R_Q)}} \|\Delta_Q \varphi\|_{L^2(\mu)} \|\Delta_R \psi\|_{L^2(\mu)}$$

and

$$\sum_{\substack{Q,R\,:\,Q\subset R,\,\ell(Q)<2^{-r}\ell(R),\\ R_Q \text{ is transit}}} |c_{R,Q}| \cdot |\langle \Delta_Q \varphi, T(\chi_{\mathbb{R}^n \setminus R_Q} b)\rangle|$$

$$\leq \frac{A\hat{A}B}{\eta(1-2^{-\frac{\tau}{2}})} \sum_{j=1}^{2^n} \sum_{Q,R\,:\,Q\subset R_j} \left[\frac{\ell(Q)}{\ell(R)}\right]^{\frac{\tau}{2}} \sqrt{\frac{\mu(Q)}{\mu(R_j)}} \|\Delta_Q \varphi\|_{L^2(\mu)} \|\Delta_R \psi\|_{L^2(\mu)}.$$

So, it is enough to demonstrate that, say, the matrix $\{T_{Q,R}\}$ defined by

$$T_{Q,R} := \left[\frac{\ell(Q)}{\ell(R)}\right]^{\frac{\tau}{2}} \sqrt{\frac{\mu(Q)}{\mu(R_1)}} \qquad (Q \subset R_1),$$

generates a bounded operator in l^2 in the sense that the following Lemma is true.

LEMMA 13.5. *For every two "sequences"* $\{a_Q\}_{Q\in\mathcal{D}_1^{tr}}$ *and* $\{b_R\}_{R\in\mathcal{D}_2^{tr}}$ *of non-negative numbers, one has*

$$\sum_{Q,R\,:\,Q\subset R_1} T_{Q,R} a_Q b_R \leq \frac{1}{1-2^{-\frac{\tau}{2}}} \Big[\sum_Q a_Q^2\Big]^{\frac{1}{2}} \Big[\sum_R b_R^2\Big]^{\frac{1}{2}}.$$

Proof. Again (as in Chapter 13.5, in Lemma 13.4) let us "slice" the matrix $T_{Q,R}$ according to the ratio $\frac{\ell(Q)}{\ell(R)}$. Namely, let

$$T_{Q,R}^{(k)} = \begin{cases} T_{Q,R}, & \text{if } Q\subset R_1,\ \ell(Q)=2^{-k}\ell(R); \\ 0, & \text{otherwise} \end{cases}$$

$(k=1,2,\dots)$. It is enough to show that for every $k\geq 0$,

$$\sum_{Q,R} T_{Q,R}^{(k)} a_Q b_R \leq 2^{-\frac{\tau}{2}k} \Big[\sum_Q a_Q^2\Big]^{\frac{1}{2}} \Big[\sum_R b_R^2\Big]^{\frac{1}{2}}.$$

The matrix $\{T_{Q,R}^{(k)}\}$ has a very good "block" structure: every a_Q can interact with *only one* variable b_R. So, it is enough to estimate each block separately, i.e., to show that for every fixed $R \in \mathcal{D}_2^{tr}$,

$$\sum_{Q\,:\,Q\subset R_1,\,\ell(Q)=2^{-k}\ell(R)} 2^{-\frac{\tau}{2}k} \sqrt{\frac{\mu(Q)}{\mu(R_1)}} a_Q b_R \leq 2^{-\frac{\tau}{2}k} \Big[\sum_Q a_Q^2\Big]^{\frac{1}{2}} b_R.$$

But, reducing both parts by the non-essential factor $2^{-\frac{\tau}{2}k} b_R$, we see that this estimate is equivalent to the trivial estimate

$$\sum_{Q\,:\,Q\subset R_1,\,\ell(Q)=2^{-k}\ell(R)} \sqrt{\frac{\mu(Q)}{\mu(R_1)}} a_Q$$

$$\leq \Big[\sum_{Q\,:\,Q\subset R_1,\,\ell(Q)=2^{-k}\ell(R)} \frac{\mu(Q)}{\mu(R_1)}\Big]^{\frac{1}{2}} \Big[\sum_Q a_Q^2\Big]^{\frac{1}{2}} \leq \Big[\sum_Q a_Q^2\Big]^{\frac{1}{2}},$$

(since cubes $Q \in \mathcal{D}_1$ of fixed size do not intersect, $\sum_{Q\,:\,Q\subset R_1,\,\ell(Q)=2^{-k}\ell(R)} \mu(Q) \leq \mu(R_1)$). $\qquad\square$

We estimated the extra terms which appear when we extend $\chi_{R_Q} b$ to the whole b. So, the extension of $\chi_{R_Q} b$ to the whole b does not cause much harm, and we get the sum

$$\sum_{\substack{Q,R:Q\subset R, \ell(Q)<2^{-r}\ell(R),\\ R_Q \text{ is transit}}} c_{R,Q}\langle \Delta_Q \varphi, Tb\rangle$$

to estimate. Note that the inner product $\langle \Delta_Q \varphi, Tb\rangle$ *does not depend* on R at all, so it seems to be a good idea to sum over R for fixed Q first. Recalling that

$$c_{R,Q} = \frac{\langle \psi\rangle_{R_Q}}{\langle b\rangle_{R_Q}} - \frac{\langle \psi\rangle_R}{\langle b\rangle_R}$$

and that $\Lambda\psi = 0 \iff \langle\psi\rangle_{R^0} = 0$, we conclude that for every $Q \in \mathcal{D}_1^{tr}$ that really appears in the above sum,

$$\sum_{\substack{R:R\supset Q, \ell(R)>2^m\ell(Q),\\ R_Q \text{ is transit}}} c_{R,Q} = \frac{\langle\psi\rangle_{R_Q}}{\langle b\rangle_{R_Q}}.$$

Definition. Let $R(Q)$ be the smallest *transit* cube $R \in \mathcal{D}_2$ containing Q and such that $\ell(R) \geq 2^r\ell(Q)$.

So, we obtain the sum

$$\sum_{Q:\ell(Q)<2^{-r}\ell(R)} \frac{\langle\psi\rangle_{R(Q)}}{\langle b\rangle_{R(Q)}} \langle\Delta_Q\varphi, Tb\rangle$$

to take care of. Let us recall that we had the convention that says that Q here are only good ones and, of course, they are only transit cubes.

The range of summation should be $Q \in \mathcal{D}_1^{tr}$, Q is good (default); there exists a cube $R \in \mathcal{D}_2^{tr}$ such that $\ell(Q) < 2^{-r}\ell(R)$, $Q \subset R$ and the child R_Q (the one containing Q) of R is transit. In other words, in fact, the sum is written formally incorrectly. We have to replace $R(Q)$ by R_Q in the summation. However, the smallest transit cube containing Q (this is $R(Q)$) and the smallest transit child (containing Q) of a certain subcube R of R^0 (this child is R_Q) are of course the one and the same cube, unless $R(Q) = R^0$. Thus the sum formally has some extra terms corresponding to $R(Q) = R^0$. But they all are zeros. In fact we work now with ψ such that $\Lambda\psi = 0$ (recall that $\Lambda\psi$ means the average of ψ with respect to μ), so $\langle\psi\rangle_{R(Q)} = 0$ if $R(Q) = R^0$.

Special paraproduct. To estimate the last sum we can ignore $\langle b\rangle_{R(Q)}$ in the denominator because $R(Q)$ are transit here, and hence the absolute values od these denominators are bounded away from zero uniformly: $|\langle b\rangle_{R(Q)}| \geq \eta$. What is left is the sum $\sum_{Q:\ell(Q)<2^{-r}\ell(R)}\langle\psi\rangle_{R(Q)}\langle\Delta_Q\varphi, Tb\rangle$.

To introduce the paraproduct operator, we rewrite our sum as follows

$$\sum_{Q:\,\ell(Q)<2^{-r}\ell(R)} \langle\psi\rangle_{R(Q)} \langle\Delta_Q\varphi, Tb\rangle = \sum_{Q:\,\ell(Q)<2^{-r}\ell(R)} \langle\psi\rangle_{R(Q)} \langle\varphi, \Delta_Q^* Tb\rangle$$

$$= \langle\varphi, \sum_{Q:\,\ell(Q)<2^{-r}\ell(R)} \langle\psi\rangle_{R(Q)} \Delta_Q^* Tb\rangle.$$

Definition. We introduce the paraproduct operator

$$\Pi_{Tb}\psi := \sum_{R\in\mathcal{D}_2,\, R\subset R^0} \langle\psi\rangle_R \sum_{Q\in\mathcal{D}_1,\, Q\text{ good and transit},\, \ell(Q)=2^{-r}\ell(R)} \Delta_Q^* Tb.$$

LEMMA 13.6. *We are under the assumptions of Theorem 12.1. For ψ such that $\langle\psi\rangle_{R^0} = 0$, we have*

$$\sum_{Q:\,\ell(Q)<2^{-r}\ell(R)} \langle\psi\rangle_{R_Q} \langle\Delta_Q\varphi, Tb\rangle = \sum_{Q:\,\ell(Q)<2^{-r}\ell(R)} \langle\psi\rangle_{R(Q)} \langle\Delta_Q\varphi, Tb\rangle = \langle\varphi, \Pi_{Tb}\psi\rangle.$$

Also,

$$(13.12) \qquad \|\Pi_{Tb}\| \leq \frac{A\,\hat{A}^2\,B\,C_6}{\eta^2}.$$

Here $B, \eta, C_6, \hat{A}$ are constants from Theorem 12.1.

Proof. Choose a test function $\varphi \in L^2(\mu)$. So, we can write

$$\langle\varphi, \Pi_{Tb}\psi\rangle = \sum_{Q:\,\ell(Q)<2^{-r}\ell(R)} \left|\langle\psi\rangle_{R(Q)} \langle\Delta_Q\varphi, Tb\rangle\right|$$

$$\leq \sum_{\substack{Q:\,\ell(Q)<2^{-r}\ell(R),\\ \|\Delta_Q\varphi\|_{L^2(\mu)}>0}} |\langle\psi\rangle_{R(Q)}| \frac{|\langle\Delta_Q\varphi, Tb\rangle|}{\|\Delta_Q\varphi\|_{L^2(\mu)}} \cdot \|\Delta_Q\varphi\|_{L^2(\mu)}$$

$$\leq \left[\sum_{\substack{Q:\,\ell(Q)<2^{-r}\ell(R),\\ \|\Delta_Q\varphi\|_{L^2(\mu)}>0}} |\langle\psi\rangle_{R(Q)}|^2 \frac{|\langle\Delta_Q\varphi, Tb\rangle|^2}{\|\Delta_Q\varphi\|_{L^2(\mu)}^2}\right]^{\frac{1}{2}} \left[\sum_Q \|\Delta_Q\varphi\|_{L^2(\mu)}^2\right]^{\frac{1}{2}}.$$

The last factor does not exceed $\frac{A\hat{A}}{\eta}\|\varphi\|_{L^2(\mu)}$ by Lemma 12.5. So, it is sufficient to show that the first factor cubed is bounded by some constant times $\|\psi\|_{L^2(\mu)}^2$. Switching to the summation over R, we see that the middle factor cubed equals

$$\sum_R |\langle\psi\rangle_R|^2 \sum_{Q\in\mathcal{F}(R)} \frac{|\langle\Delta_Q\varphi, Th\rangle|^2}{\|\Delta_Q\varphi\|_{L^2(\mu)}^2} =: \sum_R a_R |\langle\psi\rangle_R|^2,$$

where

$$\mathcal{F}(R) := \{Q : R(Q) = R, Q\text{ is transit}\}.$$

So, in order to finish the story with σ_3^{tr}, it is enough to show that the numbers a_R satisfy the Carleson condition from (12.6). Note that for every $Q \in \mathcal{F}(R)$, one has $Q \subset R$ and that the families $\mathcal{F}(R)$ are pairwise disjoint (one could say much more, but these two trivial observations are the only ones that will matter). Now, for every $S \in \mathcal{D}_2$, we have

$$\sum_{R:\, R \subset S} a_R \leq \sum_{\substack{Q:\, Q \subset S, \\ \|\Delta_Q \varphi\|_{L^2(\mu)} > 0}} \frac{|\langle \Delta_Q \varphi, Tb \rangle|^2}{\|\Delta_Q \varphi\|^2_{L^2(\mu)}} = \sum_{\substack{Q:\, Q \subset S \\ \|\Delta_Q \varphi\|_{L^2(\mu)} > 0}} \frac{|\langle \Delta_Q \varphi, \chi_S \cdot Tb \rangle|^2}{\|\Delta_Q \varphi\|^2_{L^2(\mu)}}$$

$$= \sup_{\{x_Q\} \in \ell^2(\mathcal{D}_1),\, \|\{x_Q\}\|_{\ell^2(\mathcal{D}_1)} \leq 1} \left| \sum_{\substack{Q:\, Q \subset S \\ \|\Delta_Q \varphi\|_{L^2(\mu)} > 0}} x_Q \frac{\langle \Delta_Q \varphi, \chi_S \cdot Tb \rangle}{\|\Delta_Q \varphi\|^2_{L^2(\mu)}} \right|$$

$$\leq \left\| \sum_{\substack{Q:\, Q \subset S \\ \|\Delta_Q \varphi\|_{L^2(\mu)} > 0}} x_Q \frac{\Delta_Q \varphi}{\|\Delta_Q \varphi\|} \right\|_{L^2(\mu)} \|\chi_S \cdot Tb\|_{L^2(\mu)}^2$$

$$\leq \frac{A \hat{A}^2}{\eta^2} \|\chi_S \cdot Tb\|^2_{L^2(\mu)}$$

(because of Lemma 12.5)

$$\leq \frac{A \hat{A}^2}{\eta^2} \, C_6^2 \, B \, \mu(S)$$

(because of assumption (12.2)). $\qquad\qquad\qquad\qquad\qquad\qquad\qquad\qquad\qquad\qquad \square$

CHAPTER 14

Estimate of the Diagonal Sum. Remainder in Theorem 3.3

As always, we start with a fixed $\omega = (\omega_1, \omega_2) \in \Omega^2$, thus fixing two dyadic lattices $\mathcal{D}_1$, $\mathcal{D}_2$ satisfying (7.19). For every Q we want to split $\Delta_Q \varphi$ further. Our φ was assumed to be good, so this $\Delta_Q \varphi = 0$ for bad cube Q. Splitting will be just to two zero functions in this case. To distinguish from our splitting to good and bad parts we are using subscripts g and b. Given Q (good or bad), this will be a further splitting of $\Delta_Q \varphi$ to $(\Delta_Q \varphi)_g + (\Delta_Q \varphi)_b$, where subscripts denote "better" and "worse" parts.

Recall that small parameter δ defined for us a large integer r (see (13.1)). Fix a $Q \in \mathcal{D}_1$. Consider the family of $R \in \mathcal{D}_2$ such that

$$2^{-r}\ell(Q) \le \ell(R) \le 2^r \ell(Q) \,.$$

For every such R let us introduce a fuzzy neighborhood of $sk R$:

$$fuzzy(R) := \cup_{R_i \text{ children of } R} N_{\varepsilon\ell(R)}(\partial R_i) \,,$$

where ε is a small positive number and where $N_s(\cdot)$ denotes an s-neighborhood of $(\cdot)$.

We set

$$b(Q) := Q \cap \cup_{R \in \mathcal{D}_2 : 2^{-r}\ell(Q) \le \ell(R) \le 2^r \ell(Q)} fuzzy(R) \,.$$
$$(\Delta_Q \varphi)_b := \chi_{b(Q)} \Delta_Q \varphi; \quad (\Delta_Q \varphi)_g := \Delta_Q \varphi - (\Delta_Q \varphi)_b \,.$$

We do a symmetric decomposition for $\Delta_R \psi$ for every $R \in \mathcal{D}_2$.

THEOREM 14.1. *For any $Q \in \mathcal{D}_1$,*

$$\mathbb{E}_{\omega_2}(\|(\Delta_Q \varphi)_b\|^2) \le Ar\varepsilon \|\Delta_Q \varphi\|^2 \,,$$

and for any $R \in \mathcal{D}_2$,

$$\mathbb{E}_{\omega_1}(\|(\Delta_R)_b \psi\|^2) \le Ar\varepsilon \|\Delta_Q \psi\|^2 \,,$$

Proof. Fix a point $x \in Q$. Fix integer $j \in [-r, r]$. Let p_j be a probability $\mathbb{P}_{\omega_2}$ of the event that $x \in \cup_{R \in \mathcal{D}_2 : \ell(R) = 2^j \ell(Q)} fuzzy(R)$. Relative density of the latter set (the volume of the intersection of the set with a large cube divided by the volume of the cube) is bounded by $A\varepsilon$. So $p_j \le A\varepsilon$. Hence, $\mathbb{P}_{\omega_2}\{x \in b(Q)\} = \mathbb{P}_{\omega_2}\{x \in$

$\cup_{R \in \mathcal{D}_2 : 2^{-r}\ell(Q) \leq \ell(R) \leq 2^r \ell(Q)} fuzzy(R)\} \leq A(2r+1)\varepsilon$. The rest is Fubini's theorem:

$$\begin{aligned}
\mathbb{E}_{\omega_2}(\|\Delta_Q \varphi\|^2) &= \int_E \int_\Omega \chi_{b(Q)} |\Delta_Q \varphi(x)|^2 \, d\mathbb{P}(\omega_2) d\mu(x) \\
&= \int_E \mathbb{P}_{\omega_2}\{x \in b(Q)\} |\Delta_Q \varphi(x)|^2 \, d\mu(x) \\
&\leq A(2r+1)\varepsilon \int_E |\Delta_Q \varphi(x)|^2 \, d\mu(x) = A(2r+1)\varepsilon \|\Delta_Q \psi\|^2 \, .
\end{aligned}$$

$\square$

We are ready to estimate the sum σ_1 given by

$$\sum_{Q,R:\ell(R) \geq \ell(Q) \geq 2^{-r}\ell(R), \mathrm{dist}(Q,R) \leq \ell(R)} \langle T\Delta_Q \varphi, \Delta_R \psi \rangle \, .$$

It does not matter whether Q, R are only good here; we just put an absolute value on each term and we can sum over all Q, R such that $Q \in \mathcal{D}_1$, $R \in \mathcal{D}_2$, $\ell(R) \geq \ell(Q) \geq 2^{-r}\ell(R), \mathrm{dist}(Q,R) \leq \ell(R)$. Set

$$\sigma := \sum_{Q,R:2^r\ell(R) \geq \ell(Q) \geq 2^{-r}\ell(R), \mathrm{dist}(Q,R) \leq \max(\ell(Q),\ell(R))} |\langle T\Delta_Q \varphi, \Delta_R \psi \rangle| \, .$$

As $|\sigma_1| \leq \sigma$ it is sufficient to estimate σ. Pairs Q, R as in σ will be called r-neighbors. Then

$$\sigma = \sum_{Q, R \text{ are } r\text{-neighbors}} |\langle T\Delta_Q \varphi, \Delta_R \psi \rangle| \leq \sum_{Q, R \text{ are } r\text{-neighbors}} |\langle T(\Delta_Q \varphi)_g, (\Delta_R \psi)_g \rangle|$$

$$+ \sum_{Q, R \text{ are } r\text{-neighbors}} |\langle T(\Delta_Q \varphi)_b, (\Delta_R \psi)_g \rangle| + \sum_{Q, R \text{ are } r\text{-neighbors}} |\langle T\Delta_Q \varphi, (\Delta_R \psi)_b \rangle| \, .$$

Call the terms above I, II, III. Let us estimate $\mathbb{E}II, \mathbb{E}III$:

$$\mathbb{E}II \leq \|T\|\mathbb{E}\Big(\sum_{(Q,R) \text{ are } r\text{-neighbors}} \|(\Delta_Q \varphi)_b\| \|(\Delta_R \psi)_g\| \Big) \, .$$

$$\mathbb{E}III \leq \|T\|\mathbb{E}\Big(\sum_{(Q,R) \text{ are } r\text{-neighbors}} \|\Delta_Q \varphi\| \|(\Delta_R \psi)_b\| \Big) \, .$$

We denote

$$(14.1) \qquad R_2(\omega, \varphi, \psi) := \sum_{(Q,R) \text{ are } r\text{-neighbors}} \|(\Delta_Q \varphi)_b\| \|\Delta_R \psi\| \, .$$

$$(14.2) \qquad R_3(\omega, \varphi, \psi) := \sum_{(Q,R) \text{ are } r\text{-neighbors}} \|\Delta_Q \varphi\| \|(\Delta_R \psi)_b\| \, .$$

Let us estimate $\mathbb{E}R_2(\omega, \varphi, \psi)$ by

$$\Big(\sum_{(Q,R) \text{ are } r\text{-neighbors}} \mathbb{E}\|(\Delta_Q \varphi)_b\|^2 \Big)^{1/2} \Big(\sum_{(Q,R) \text{ are } r\text{-neighbors}} \mathbb{E}\|(\Delta_R \psi)_g\|^2 \Big)^{1/2} \, .$$

But for any $Q \in \mathcal{D}_1$ there are at most 2^{Ar} cubes $R \in \mathcal{D}_2$ such that (Q, R) are r-neighbors. So

$$\mathbb{E}R_2(\omega, \varphi, \psi) \leq 2^{Ar} \Big(\sum_{Q \in \mathcal{D}_1} \mathbb{E}\|(\Delta_Q \varphi)_b\|^2 \Big)^{1/2} \Big(\sum_{R \in \mathcal{D}_2} \mathbb{E}\|(\Delta_R \psi)_g\|^2 \Big)^{1/2}$$

$$= 2^{Ar} \Big(\sum_{Q \in \mathcal{D}_1} \mathbb{E}_{\omega_1} \mathbb{E}_{\omega_2} \|(\Delta_Q \varphi)_b\|^2 \Big)^{1/2} \Big(\sum_{R \in \mathcal{D}_2} \mathbb{E}_{\omega_2} \mathbb{E}_{\omega_1} \|(\Delta_R \psi)_g\|^2 \Big)^{1/2} .$$

It is time to use Theorem 14.1 to get

$$\mathbb{E}R_2(\omega, \varphi, \psi) \leq A(2r+1)2^{Ar} \varepsilon \Big(\sum_{Q \in \mathcal{D}_1} \|\Delta_Q \varphi\|^2 \Big)^{1/2} \Big(\sum_{R \in \mathcal{D}_2} \mathbb{E}_{\omega_2} \|\Delta_R \psi\|^2 \Big)^{1/2} .$$

Using Lemma 12.5 we get

(14.3) $$\mathbb{E}R_2(\omega, \varphi, \psi) \leq A(2r+1)2^{Ar} C(\hat{A}, \eta) \varepsilon \|f\| \|g\| .$$

Similarly,

(14.4) $$\mathbb{E}R_3(\omega, \varphi, \psi) \leq A(2r+1)2^{Ar} C(\hat{A}, \eta) \varepsilon \|f\| \|g\| .$$

We are left to estimate

$$I := \sum_{(Q, R) \text{ are } r\text{-neighbors}} |\langle T(\Delta_Q \varphi)_g, (\Delta_R \psi)_g \rangle| .$$

Let us split it to $\Sigma_{term} + \Sigma_{tr}$, where

$$\Sigma_{tr} := \sum_{(Q, R) \text{ are } r\text{-neighbors, all children of } Q, R \text{ are transit}} |\langle T(\Delta_Q \varphi)_g, (\Delta_R \psi)_g \rangle| .$$

The summation in Σ_{term} goes over r-neighbors (Q, R) such that there exists a child of Q or R that is terminal.

14.1. Estimate of Σ_{term}

Let Q, R be as in Σ_{term}. Recall that skQ is the union of boundaries of children of Q. Then $skR \cup skQ$ split $Q \cup R$ into disjoint cells. We combine some of them into two new cells $Q \setminus R$ and $R \setminus Q$. Each of the other cells is the intersection of some child of Q with a child of R. The number of these cells is at most $N = N(n) < \infty$. Call them $\{C_k\}_{k=1}^N$. Set

$$\varphi_k := \chi_{C_k} (\Delta_Q)_g \varphi, \ \psi_k := \chi_{C_k} (\Delta_R)_g \psi .$$

These are restrictions of $(\Delta_Q)_g \varphi$, and $(\Delta_R)_g \psi$ on cells. Some of these φ_k, ψ_k can identically vanish. The next estimate involves our ubiquitous integer r, which comes into our definition of r-neighbors (see (13.1) about it). We will prove now that (for all $k, j + 1, \ldots, N$)

(14.5) $$|\langle T\varphi_k, \psi_j \rangle| \leq \frac{AC 2^{Amr}}{\varepsilon^m} \|\Delta_Q \varphi\| \|\Delta_R \psi\|$$
$$+ \|T\| \|(\Delta_Q \varphi)_b\| \|\Delta_Q \psi\| + \|T\| \|\Delta_Q \varphi\| \|(\Delta_Q \psi)_b\| ,$$

where C is a finite constant depending only on constants of Theorem 12.1. The estimate 14.5 is sufficient to obtain (for the definition of $R_2(\omega, \varphi, \psi), R_3(\omega, \varphi, \psi)$, see (14.1), (14.2))

(14.6) $$\Sigma_{term} \leq \frac{2^{Amr} AC}{\eta^2 \varepsilon^m} \|f\| \|g\| + A(R_2(\omega, \varphi, \psi) + R_3(\omega, \varphi, \psi)) .$$

Indeed, 14.5 implies (here Q or R or both have terminal children and are r-neighbors)

$$|\langle T(\Delta_Q\varphi)_g, (\Delta_R\psi)_g\rangle| \leq \frac{N^2 2^{Amr}AC}{\varepsilon^m}\|\Delta_Q\varphi\|\|\Delta_R\psi\|$$
$$+ N^2\|T\|\|(\Delta_Q\varphi)_b\|\|\Delta_Q\psi\| + N^2\|T\|\|\Delta_Q\varphi\|\|(\Delta_Q\psi)_b\|.$$

Using the convention that all finite constants depending only on the dimension n are called A, we continue (notice that in the summation we "forget" the fact about terminal children, we add maybe some terms, and sum over all pairs of r-neighbors)

$$\Sigma_{term} \leq \sum_{(Q,R) \text{ are } r\text{-neighbors}} |\langle T(\Delta_Q\varphi)_g, (\Delta_R\psi)_g\rangle| + AR_2(\omega,\varphi,\psi) + AR_3(\omega,\varphi,\psi)$$

$$\leq \sum_{(Q,R) \text{ are } r\text{-neighbors}} \frac{2^{Amr}AC}{\varepsilon^m}\|(\Delta_Q\varphi)_g\|\|(\Delta_R\psi)_g\|$$
$$+ AR_2(\omega,\varphi,\psi) + AR_3(\omega,\varphi,\psi)$$

$$\leq \frac{2^{Amr}AC}{\varepsilon^m}\Big(\sum_{(Q,R) \text{ are } r\text{-neighbors}} \|\Delta_Q\varphi\|^2\Big)^{1/2}\Big(\sum_{(Q,R) \text{ are } r\text{-neighbors}} \|\Delta_R\psi\|^2\Big)^{1/2}$$
$$+ AR_2(\omega,\varphi,\psi) + AR_3(\omega,\varphi,\psi)$$

$$\leq \frac{2^{Amr}AC}{\varepsilon^m}\Big(\sum_{Q\in\mathcal{D}_1}\|\Delta_Q\varphi\|^2\Big)^{1/2}\Big(\sum_{R\in\mathcal{D}_2}\|\Delta_R\psi\|^2\Big)^{1/2}$$
$$+ AR_2(\omega,\varphi,\psi) + AR_3(\omega,\varphi,\psi).$$

Here we used the fact that each $Q \in \mathcal{D}_1$ has at most 2^{Ar} r-neighbors $R \in \mathcal{D}_2$ (symmetrically for $R \in \mathcal{D}_2$). To estimate the last sums we use Lemma 12.5. This is how η comes into the denominator of (14.6).

The estimate (14.6) of Σ_{term} will be completed when we prove (14.5).

The proof of (14.5). Consider first the case $k \neq j$. Then use the fact

$$(14.7) \qquad \text{dist}(\text{supp}\,\varphi_k, \text{supp}\,\psi_j) \geq 2^{-r}\ell(Q)\varepsilon.$$

This is just because φ_k, ψ_j are restrictions of $(\Delta_Q)_g\varphi, (\Delta_R)_g\psi$ to disjoint cells and the fact that supports of $(\Delta_Q)_g\varphi, (\Delta_R)_g\psi$ do not intersect $fuzzy(Q), fuzzy(R)$ correspondingly. Using (14.7) we write the rough estimate on the kernel $k(x,y)$:

$$x \in \text{supp}\,\varphi_k, y \in \text{supp}\,\psi_j \Rightarrow |k(x,y)| \leq \frac{C}{|x-y|^m} \leq \frac{C2^{mr}}{\ell(Q)^m\varepsilon^m}.$$

Hence,

$$(14.8) \qquad |\langle T\varphi_k, \psi_j\rangle| \leq \frac{C2^{mr}}{\ell(Q)^m\varepsilon^m}\|f_k\|_{L^1}\|\psi_j\|_{L^1}.$$

From here,

$$|\langle T\varphi_k, \psi_j\rangle| \leq \frac{C2^{mr}}{\ell(Q)^m\varepsilon^m}\|f_k\|\|\psi_j\|(\mu(C_k))^{1/2}(\mu(C_j))^{1/2}.$$

Now $\|f_k\| \leq \|\Delta_Q\varphi\|, \|\psi_j\| \leq \|\Delta_R\psi\|$, and we also have

$$\mu(C_k) + \mu(C_j) \leq \mu(Q) + \mu(R) \leq AB(\text{diam}(Q)^m + \text{diam}(R)^m) \leq AB2^{Amr}\ell(Q)^m.$$

The last inequality is clear because Q, R are r-neighbors. The one before the last follows from the fact that Q, R are *transit* cubes and for them we have Lemma 12.4. Gathering the last estimates together, we obtain

$$|\langle T\varphi_k, \psi_j \rangle| \le \frac{C2^{Amr}}{\varepsilon^m} \|\Delta_Q \varphi\| \|\Delta_R \psi\|,$$

which is (14.5) for $k \ne j$.

The estimate of $|\langle T\varphi_k, \psi_k \rangle|$ when C_k lies in transit children of both Q and R. To estimate $|\langle T\varphi_k, \psi_k \rangle|$, let us assume now that $C_k = Q_s \cap R_t$ (where Q_s is a child of Q, R_t is a child of R), and suppose that Q_s and R_t are transit. We cannot use strong disjointedness of supports of φ_k, ψ_j as in (14.7), because the supports can intersect. We cannot use the rough bound on k: $|k(x,y)| \le \frac{C}{|x-y|^m}$, because now it is useless: x can be as close to y as one wishes. But if Q_s and R_t are transit and $C_k = Q_s \cap R_t$, then

(14.9) $$\Delta_Q \varphi | C_k = c_Q b, \quad \Delta_R \psi | C_R = c_R b,$$

where c_Q, c_R are two constants. This is just the nature of our decomposition (see Section 12.2). We want to show

(14.10) $$|\langle T\varphi_k, \psi_k \rangle| \le \|T\| \|(\Delta_Q \varphi)_b\| \|\Delta_Q \psi\| + \|T\| \|\Delta_Q \varphi\| \|(\Delta_Q \psi)_b\|.$$

Of course then (see the definition of $R_2(\omega, \varphi, \psi), R_3(\omega, \varphi, \psi)$ in (14.1), (14.2))

(14.11) $$\sum_{(Q,R) \text{ } r\text{-neighbors}} \sum_{k: C_k = Q_s \cap R_t, \text{ } Q_s, R_t \text{ transit}} |\langle T\varphi_k, \psi_k \rangle|$$
$$\le \|T\| (R_2(\omega, \varphi, \psi) + R_3(\omega, \varphi, \psi)).$$

The estimate of the average then follows from (14.3), (14.4).

We are left to prove (14.10). Along with $\varphi_k = \chi_{C_k}(\Delta_Q \varphi)_g$, consider $f_k := \chi_{C_k}(\Delta_Q \varphi)_b$ and along with $\psi_k = \chi_{C_k}(\Delta_Q \psi)_g$, consider $g_k := \chi_{C_k}(\Delta_Q \psi)_b$. Then
$$|\langle T\varphi_k, \psi_k \rangle| \le |\langle T(\varphi_k + f_k), (\psi_k + g_k) \rangle| + |\langle T(\varphi_k + f_k), g_k \rangle| + |\langle T f_k, \psi_k \rangle| =: I + II + III.$$
But
$$II = |\langle \chi_{C_k} \Delta_Q \varphi, \chi_{C_k}(\Delta_R \psi)_b \rangle| \le \|T\| \|\Delta_Q \varphi\| \|(\Delta_Q \psi)_b\|,$$
and
$$III = |\langle T f_k, \psi_k \rangle| = |\langle \chi_{C_k}(\Delta_Q \varphi)_b, \chi_{C_k}(\Delta_R \psi)_g \rangle| \le \|T\| \|(\Delta_Q \varphi)_b\| \|\Delta_Q \psi\|.$$
To prove (14.10) we are left to estimate I. But by (14.9)
$$I = |\langle T\chi_{C_k} \Delta_Q \varphi, \chi_{C_k} \Delta_R \psi \rangle| = |c_Q| |c_R| |\langle T\chi_{C_k} b, \chi_{C_k} b \rangle| = 0$$
by the antisymmetry of T.

The estimate of $|\langle T\varphi_k, \psi_k \rangle|$ when C_k lies in a terminal child of Q or R. To complete the estimate of expressions $|\langle T\varphi_k, \psi_k \rangle|$, let us assume now that $C_k = Q_s \cap R_t$ (where Q_s is a child of Q, R_t is a child of R), and suppose that R_t is terminal. We cannot use strong disjointedness of supports of φ_k, ψ_j as in (14.7), because the supports can intersect. We cannot use the rough bound on k: $|k(x,y)| \le \frac{C}{|x-y|^m}$, because now it is useless: x can be as close to y as one wishes. But because of the fact that R_t is terminal we can use Lemma 12.3; namely, (12.9) gives us

(14.12) $$x, y \in C_k \Rightarrow |k(x,y)| \le \min[\frac{1}{(\text{dist}(x, \partial R_t))^m}, \frac{1}{(\text{dist}(y, \partial R_t))^m}].$$

But by construction of $\Delta_Q\varphi$ we have

(14.13) $\varphi_k(x) = \Delta_Q\varphi(x) = 0, \ \forall x \in C_k : \text{dist}(x, \partial R_t) \le \ell(R)\varepsilon \ge 2^{-r}\ell(Q)\varepsilon\,.$

Now we use (14.12) and (14.13) together to conclude that

$$\int\int |k(x,y)||\varphi_k(x)||\psi_k(y)|\, d\mu(x)d\mu(y) \le \frac{2^{mr}}{\ell(Q)^m \varepsilon^m} \|f_k\|_{L^1}\|\psi_k\|_{L^1}\,.$$

We had this estimate already; see (14.8). From it we conclude, exactly as above, that

$$|\langle T\varphi_k, \psi_k\rangle| \le \frac{C2^{Amr}}{\varepsilon^m}\|\Delta_Q\varphi\|\|\Delta_R\psi\|\,,$$

which is (14.5) for $k = j$ if C_k lies in a terminal child of R. Symmetrically, the same holds for cells in the terminal child of Q.

14.2. Estimate of Σ_{tr}

This has already been done. We repeat verbatim the previous section, but without the estimate of $|\langle T\varphi_k, \psi_k\rangle|$ when C_k lies in a terminal child of Q or R. In brief, we have already proved

(14.14) $\Sigma_{tr} \le \frac{2^{Ar}AC}{\eta^2\varepsilon^m}\|f\|\|g\| + A(R_2(\omega,\varphi,\psi) + R_3(\omega,\varphi,\psi))\,.$

This is it. The last reduction, namely Theorem 10.6, is completely proved.

As we have shown this proves Theorems 9.1, 8.1, 7.1, and 6.1 in that order. All of the reductions have been made in corresponding chapters above.

CHAPTER 15

Two–Weight Estimate for the Hilbert Transform. Preliminaries

We start now the second part of our exposition. This second part is devoted to a seemingly very different subject. The reason we consider this subject—two–weight estimates for some Calderón–Zygmund operators—is that the technique developed above for degenerate (nonhomogeneous) cases of Tb theorem (and used above to solve Vitushkin's conjecture and its relatives), seems to work very well also for this other quite intriguing problem from the theory of Calderón–Zygmund operators.

Let us recall a little bit of the history of the problem. We will be mentioning only the Hilbert transform—the common model of a Calderón–Zygmund operator. In 1960 Helson and Szegö in [**24**] described the weights such that, say, for all f smooth with compact support on the real line $\mathbb{R}^n$,

$$(15.1) \qquad \int_{\mathbb{R}} |Hf|^2 \, wdx \le C \int_{\mathbb{R}} |f|^2 \, wdx \,,$$

where the Hilbert transform H is defined as follows:

$$(15.2) \qquad Hf(x) := \frac{1}{\pi} p.v. \int_{\mathbb{R}} \frac{f(t)}{x-t} \, dt := \lim_{\varepsilon \to 0+} \frac{1}{\pi} \int_{t:|t-x|\ge\varepsilon} \frac{f(t)}{x-t} \, dt \,.$$

Here is the description of Helson and Szegö: the weight satisfies (15.1) if and only if

$$(15.3) \qquad \log w = u + Hv, \ u,v \in L^\infty, \ \|v\|_\infty < \frac{\pi}{2} \,.$$

In 1971 a new description of such weights appeared. This description was due to Hunt, Muckenhaupt and Wheeden [**23**], and it was in totally different terms:

$$(15.4) \qquad Q_w := \sup_{I \subset \mathbb{R}} \langle w \rangle_I \langle w^{-1} \rangle_I < \infty \,.$$

Here I run over all finite intervals of the real line. Note that so far there is no direct proof that (15.4) implies (15.3). Of course the problem with two weights attracted the attention. The problem is to describe the pairs of nonzero weights such that

$$(15.5) \qquad \int_{\mathbb{R}} |HF|^2 \, vdx \le C \int_{\mathbb{R}} |F|^2 \, udx \,.$$

There is a vast amount of literature about the two–weight problems. Now we mention only the works of P. Koosis [**29**], [**30**].

One–weight inequality became very important because of its relations with the theory of Toeplitz operators and with the spectral theory of stationary stochastic processes; see [**53**],[**66**], [**67**].

Two–weight inequality first attracted the attention because of its obvious relation to the one–weight counterpart. But recently it became clear that it is very essential in perturbation theory of unitary and self-adjoint operators. In particular, the question, when a rank one perturbation of a unitary operator is similar to a unitary operator, is essentially about the two–weight estimate of the Hilbert transform; see [**40**], for example. Subtle questions about the subspaces of the Hardy class H^2 invariant under the inverse shift operators also are essentially about the two–weight Hilbert transform; see [**42**]. And at last, recently P. Yuditsky indicated to me how the two–weight Hilbert transform appears naturally in certain unsolved questions concerning the orthogonal polynomials.

Let us formulate the two–weight Hilbert transform problem in form that is more convenient to us than (15.5). Let μ, ν be two positive measures on $\mathbb{R}$. We define the Hilbert transform H_μ from $L^2(\mu)$ to $L^2(\nu)$ as any bounded linear operator from $L^2(\mu)$ to $L^2(\nu)$ such that

$$(15.6) \qquad H_\mu f(x) := \frac{1}{\pi} \int_{\mathbb{R}} \frac{f(t)}{x-t} \, d\mu(t), \; \forall x \in \mathbb{R} \setminus \mathrm{supp}(f) \,.$$

Such an operator is not uniquely defined, because it defines $H_\mu f$ only outside of the support of f. But we will prove the main result for all such operators. Notice that the adjoint H_μ^* is just $-H_\nu$; it is also just a Hilbert transform in our sense (up to a minus sign).

In what follows we always consider only the measures without atoms.

Let us change the variables in (15.5): $d\mu := \frac{1}{u} \, dx, F := \frac{f}{u}, d\nu := v dx$. Then (15.5) transforms itself into

$$(15.7) \qquad \int_{\mathbb{R}} |H_\mu f|^2 \, d\nu \leq C \int_{\mathbb{R}} |f|^2 \, d\mu \,.$$

A very subtle point is that we are not interested when (15.7) holds with the same finite C for all f in $L^2(\mu)$. We already assumed by definition that this is the case. What we are interested in are some simple characteristics computable by means of μ and ν, and such that C can be estimated by these characteristics. An example of one such characteristic is

$$(15.8) \qquad Q_{\mu,\nu} := \sup_{I \subset \mathbb{R}} \langle \mu \rangle_I \langle \nu \rangle_I := \sup_{I \subset \mathbb{R}} \frac{\mu(I)}{|I|} \frac{\nu(I)}{|I|} \,.$$

This is a total analog of Q_w from (15.4). In fact, in a one–weight case $u = v = w$ of (15.5), we have $d\mu = \frac{1}{w} \, dx, d\nu = w dx$, and so $Q_{\mu,\nu}$ becomes Q_w. We will see soon that

$$(15.9) \qquad Q_{\mu,\nu}^{1/2} \leq A \|H_\mu\|_{L^2(\mu) \to L^2(\nu)} \,.$$

Of course, we are interested in a sort of opposite estimate. After all, the Hunt–Muckenhoupt–Wheeden theorem from [**23**] says that the finiteness of Q_w is equivalent to the boundedness of the corresponding Hilbert transform. Moreover, recently S. Petermichl [**54**] proved that

$$\|H_\mu\|_{L^2(\mu) \to L^2(\nu)} \leq A \, Q_{\mu,\nu} = A \, Q_w$$

in a one–weight case, that is when $d\mu = \frac{1}{w} \, dx, d\nu = w dx$. See also [**56**], where this is proved for the Ahlfors–Beurling transform instead of the Hilbert transform.

However, in a two–weight case, nothing like the full analog of [**23**]'s result is possible. Not to say that [**54**] or [**56**] can be proved here. Strangely enough, this has been understood only recently due to the work of F. Nazarov [**41**]. See also [**40**], [**42**].

At any rate $Q_{\mu,\nu}$ will be an important characteristic of "Hunt-Muckenhoupt–Wheeden" type, which will play an important part in estimating

$$\|H_\mu\|_{L^2(\mu)\to L^2(\nu)}.$$

The only thing we have said is that it alone is not sufficient. One has to look for other μ, ν-quantities.

It is important to mention that unlike the "Hunt–Muckenhoupt–Wheeden" type characteristics of two–measure (two–weight), the "Helson–Szegő" type characteristics were found long ago. This has been done in the papers of Cotlar and Sadosky [7]-[8]. Paper [6] gives another equivalence to the Helson–Szegő condition. Papers [9]–[10] also treat the Helson–Szegő type theorem in L^p for the case $p \neq 2$.

From what we have described above it becomes clear that we are after "Hunt–Muckenhoupt–Wheeden" type characteristics of two–measure (two–weight), which, together with characteristic $Q_{\mu,\nu}$, will allow us to estimate $\|H_\mu\|_{L^2(\mu)\to L^2(\nu)}$.

The difficulty is twofold. First of all, two–weight problems have a huge degree of freedom in comparison to rather rigid one–weight problems. This is why one quantity $Q_{\mu,\nu}$ is not sufficient. Secondly we are dealing with a singular operator. Singular kernels are much more difficult to deal with than are positive kernels. Maybe for operators with positive kernels the two–weight problems are more easily approachable? It has been found in the mid 80's that this is the case.

E. Sawyer was the first who fully characterized the boundedness of several important operators with positive kernels between two different weighted spaces. This concerned in particular the maximal operator and the Carleson imbedding theorem. The reader is referred to [58], [59], [28] and also to [50], where Sawyer's results got Bellman function explanation and interpretation. Sawyer's conditions were simple and beautiful; they were, in a sense, of "Hunt–Muckenhoupt–Wheeden" type. But actually their meaning was very transparent:

> often operator with positive kernel between two weighted L^2-spaces
>
> is bounded if and only if it is uniformly bounded on a system
>
> of simple test functions and the same holds for its adjoint.

It is usually enough to take the characteristic functions of the intervals (cubes) as the family of test functions.

This was a remarkable discovery. Actually, almost at the same time, a series of works by G. David and J.-L. Journé appeared, devoted to the so-called $T1$ theorems. Here the main object was singular operators (kernel changes the sign), more precisely, Calderón–Zygmund operators. The answer (these $T1$ theorems) was in the same spirit: check T and T^* on characteristic functions of intervals. But unlike the case considered by Sawyer, these problems of David and Journé were **one–weight problems** in the following sense: given the operator with Calderón–Zygmund kernel k, bounded in $L^2(\mu)$, one looks for characteristics, which allow one to estimate the norm of this operator. The phrase "given the operator T with Calderón–Zygmund kernel k" means that we are given a positive measure μ (say in $\mathbb{R}^n$) and

$$(15.10) \qquad T_\mu f(x) := \int_{\mathbb{R}^n} k(x,t)f(t)\,d\mu(t) \ \ \forall x \in \mathbb{R}^n \setminus \mathrm{supp}(f)\,.$$

There are many such operators, of course. But David–Journé were looking for characteristics which give the bound on the norms of all such operators, meaning the norms from $L^2(\mu)$ to the same $L^2(\mu)$. This is why we call such problems one–weight problems; they concern the estimate of $\|T_\mu : L^2(\mu) \to L^2(\mu)\|$. Notice that here one–weight problem means something quite different than in the Hunt-Muckenhoupt-Wheeden theorem. In the Hunt-Muckenhoupt-Wheeden theorem, one deals with $\|H_\mu : L^2(\mu) \to L^2(\nu)\|$, but for a very special case: $\nu = wdx, \mu = \frac{1}{w}dx$.

The last important remark is that the theory of David-Journé (usually united under the name "$T1$ theorems") originally concerned only one measure μ, namely, Lebesgue measure in $\mathbb{R}^n$: $d\mu = dx$. It was noticed that for doubling measures one can construct a series of $T1$ theorems. This has been done in a paper by M. Christ [5]. The doubling property seemed to be a cornerstone of David–Journé–Christ theory of Calderón–Zygmund operators.

However, a strong need to get rid of this cornerstone appeared from the attempt to solve Vitushkin's problems. See sections above, where the exposition is made of nondoubling (nonhomogeneous) theory of Calderón–Zygmund operators.

Summarizing all of this: one–weight problems (in both senses indicated above) are difficult, but basically solved for both Calderón–Zygmund operators and for the operators with positive kernels.

Two–weight problems are considerably more difficult, but basically solved for the wide class of operators with positive kernels.

We consider the worst of both worlds. Our operators will be singular (here we consider just a model, the Hilbert transform) and instead of a one–weight problem we consider a two–weight problem. This is why we need all the tricks from the sections above dealing with nonhomogeneous $T1$ and Tb theorems.

Here is our main result concerning two–weight Hilbert transform. It uses fully the box of tools we applied above to construct a nonhomogeneous version of Calderón–Zygmund theory (criteria for the boundedness of Calderón–Zygmund $T_\mu : L^2(\mu) \to L^2(\mu)$ for nondoubling μ). The huge drawback of what will follow is that we will be obliged to impose the doubling conditions on μ, ν if we want to prove a simple Hunt–Muckenhoupt–Wheeden (actually Sawyer) type result on boundedness of $\|H_\mu : L^2(\mu) \to L^2(\nu)\|$. This unwelcome but persistent doubling assumption is probably not needed: the result should be true in general. But the huge difficulty of a two–weight estimate for singular operators forced us to impose this assumption. This is especially strange because we use the "nonhomogeneous" technique, which is supposed to smoothen up all degeneracies of the measures; and it does, as sections above show, but so far only for one–weight problems.

Assumption. μ, ν are measures on $\mathbb{R}$ such that (the symbol I stands for the interval, as usual)

(15.11) $\operatorname{supp}\mu = \operatorname{supp}\nu = \mathbb{R}$, and $\mu(2I) \leq C_d\mu(I),\ \nu(2I) \leq C_d\nu(I),\ \forall I \subset \mathbb{R}$.

Let us recall that we introduced $Q_{\mu,\nu}$ in (15.8). Its finiteness is necessary for the boundedness of the corresponding two–weight Hilbert transform. We will see this soon. But actually there is a slightly larger quantity, more convenient for us. Its finiteness is necessary for the boundedness of the corresponding two–weight Hilbert transform too. Let us introduce it. Recall that Poisson extension of measure supported by $\mathbb{R}$ is given by the formula

$$P_\mu(z) := \frac{1}{\pi} \int_{\mathbb{R}} \frac{\Im z}{(\Re z - t)^2 + (\Im z)^2}\, dt\,.$$

Put

(15.12)
$$PQ_{\mu,\nu} := \sup_{z \in \mathbb{C}_+} P_\mu(z) P_\nu(z).$$

It is easy to see that there exists an absolute constant A such that for any pair of positive measures,

(15.13)
$$Q_{\mu,\nu} \leq A\, PQ_{\mu,\nu}.$$

In what follows we always consider only the measures without atoms.

THEOREM 15.1. *Let* μ, ν *be as in* (15.11) *with doubling constant* $C_d < \infty$. *Let* H_μ, H_ν *be bounded on characteristic functions, namely*

(15.14)
$$\|H_\mu \chi_I\|_{L^2(\nu)} \leq C_\chi \nu(I), \quad \forall I \subset \mathbb{R},$$

(15.15)
$$\|H_\nu \chi_I\|_{L^2(\mu)} \leq C_\chi \mu(I), \quad \forall I \subset \mathbb{R}.$$

Let also

(15.16)
$$PQ_{\mu,\nu} = \sup_{z \in \mathbb{C}_+} P_\mu(z) P_\nu(z) \leq C_p.$$

Then $\|H_\mu : L^2(\mu) \to L^2(\nu)\|$ *is bounded by constant* $C(C_d, C_\chi, C_p) < \infty$, *which depends only on* C_d, C_χ, C_p.

We can call it "Sawyer's theorem for the Hilbert transform". We hope that (15.11) is superfluous.

CHAPTER 16

Necessity in the Main Theorem

Assumptions (15.14), (15.15) are obviously necessary. As to (15.16), it is necessary as well. In fact, let us consider (just for the sake of convenience) our measures μ, ν on the unit circle $\mathbb{T}$ (instead of being on the line).

As in (15.6), we define the two–weight Hilbert transform on the circle as follows. Let μ, ν be two positive and *finite* measures on $\mathbb{T}$. We define the Hilbert transform H_μ from $L^2(\mu)$ to $L^2(\nu)$ as *any bounded linear operator* from $L^2(\mu)$ to $L^2(\nu)$ such that

$$(16.1) \qquad H_\mu f(x) := \frac{1}{2\pi} \int_\mathbb{T} \frac{f(\zeta)}{1 - \bar{\zeta}z} \, d\mu(\zeta), \; \forall x \in \mathbb{T} \setminus \operatorname{supp}(f) \,.$$

We recall that the Poisson integral of the measure on $\mathbb{T}$ is given by

$$(16.2) \qquad P_\mu(a) := \frac{1}{2\pi} \int_\mathbb{T} \frac{1 - |a|^2}{|1 - \bar{a}z|^2} \, d\mu(z), \; a \in \mathbb{D} \,.$$

In what follows, we always consider only the measures without atoms. Here is the explanation. We want to get the necessity of (15.16). Suppose μ, ν are both delta measures at the same point. But we adopted such a definition of H_μ, which allows for its non-uniqueness. Two H_μ may differ by the bounded operator from $L^2(\mu)$ to $L^2(\nu)$ that preserves the support of a function, that is by the operator of multiplication. In particular, in the case when $\mu = \nu = \delta_1$, we can see that the identity operator is also H_μ. But $PQ_{\mu,\nu} = \infty$ obviously.

For a point $a \in \mathbb{D}$ put $b_a(z) := \frac{z-a}{1-\bar{a}z}$. This is a Blaschke factor, and so it is a unimodular on the circle. So the operator M_{b_a} of multiplication on b_a is an isometry in any $L^2(\sigma), \operatorname{supp} \sigma \subset \mathbb{T}$. Given a bounded operator $H_\mu : L^2(\mu) \to L^2(\nu)$, consider a new operator given by

$$T_{\mu,a} := H_\mu - M_{\bar{b}_a} H_\mu M_{b_a} \,.$$

Then (16.1) implies that

$$(16.3) \qquad T_{\mu,a}f(z) = \frac{1}{2\pi} \int_\mathbb{T} \frac{1 - \overline{b_a(\zeta)}b_a(z)}{1 - \bar{\zeta}z} f(\zeta) \, d\mu(\zeta), \; \forall z \in \mathbb{T} \setminus \operatorname{supp}(f) \,.$$

An easy computation shows

$$(16.4) \qquad \forall \zeta, z \in \mathbb{T}, \forall a \in \mathbb{D}, \quad \frac{1 - \overline{b_a(\zeta)}b_a(z)}{1 - \bar{\zeta}z} = \frac{1 - |a|^2}{(1 - a\bar{\zeta})(1 - \bar{a}z)} \,.$$

In particular, the kernel in (16.3) is bounded.

Let us present the idea of the rest of the proof. The norm of such an operator (as an operator from $L^2(\mu)$ to $L^2(\nu)$) should be of course just $\|k_a\|_\mu \|k_a\|_\nu$, which is obviously (see (16.2)) $(P_\mu(a)P_\nu(a))^{1/2}$. On the other hand

$$(16.5) \qquad \|T_{\mu,a}\| = \|H_\mu - M_{\bar{b}_a}H_\mu M_{b_a}\| \le 2\|H_\mu\|,$$

as multiplications on $b_a, \bar{b}_a$ are isometries in $L^2(\mu), L^2(\nu)$.

Combining with (16.5), one gets

$$(16.6) \qquad (P_\mu(a)P_\nu(a))^{1/2} \le 2\|H_\mu\|.$$

So we would get (15.16).

The problem with the "proof" above is that the operator T_a with the kernel

$$\frac{1 - |a|^2}{(1 - a\bar{\zeta})(1 - \bar{a}z)},$$

and the operator $T_{\mu,a}$ may be different. In fact, (16.3) says only that

$$(T_{\mu,a}f, g)_\nu = (T_a f, g)_\nu, \ \forall f \in L^2(\mu), g \in L^2(\nu), \ \mathrm{supp}(f) \cap \mathrm{supp}(g) = \emptyset.$$

From this alone we do not get (16.6), but we get only its weaker version:

$$(16.7) \qquad (P_{\mu|E}(a)P_{\nu|F}(a))^{1/2} \le 2\|H_\mu\|, \ \forall E, F \subset \mathbb{T}, E \cap F = \emptyset.$$

We are left to explain why (16.7) implies (16.6). They are both Möbius invariant, and so let $a = 0$. As μ has no atoms we can choose E_1 to be a half-circle such that $\mu(E_1) = \frac{1}{2}\mu(\mathbb{T})$. Let $E_2 = \mathbb{T} \setminus E_1$. Call F such an E_i that has larger ν measure. The other one is called E. For example, if $\nu(E_1) \ge \nu(E_2)$, we have $F = E_1, E = E_2$. Then of course, $P_\mu(0)P_\nu(0) \le 4\, P_{\mu|E}(0)P_{\nu|F}(0)$. And (16.6) follows from (16.7).

CHAPTER 17

Two–Weight Hilbert Transform. Towards the Main Theorem

In what follows we use Nazarov–Treil–Volberg's paper [**48**]. Nazarov also noticed that what follows can be used for a wide class of Calderón–Zygmund operators.

Let $f \in L^2(\mu), g \in L^2(\nu)$ be two test functions. We can think without the loss of generality that they have compact support. Then let us think that their support is in $[\frac{1}{4}, \frac{3}{4}]$. Let $\mathcal{D}^\mu, \mathcal{D}^\nu$ be two dyadic lattices of $\mathbb{R}$. We can think that they are both shifts of the same standard dyadic lattice $\mathcal{D}$, such that $[0, 1] \in \mathcal{D}$, and that $\mathcal{D}^\mu = \mathcal{D} + \omega_1, \mathcal{D}^\nu = \mathcal{D} + \omega_2$, where $\omega_1, \omega_2 \in [-\frac{1}{4}, \frac{1}{4}]$. We have a natural probability space of pairs of such dyadic lattices:

$$\Omega := \{(\omega_1, \omega_2) \in [-\frac{1}{4}, \frac{1}{4}]^2\}$$

provided with probability $\mathbb{P}$ which is equal to normalized Lebesgue measure on $[-\frac{1}{4}, \frac{1}{4}]^2$. We called these two independent dyadic lattices $\mathcal{D}^\mu, \mathcal{D}^\nu$ because they will be used to decompose $f \in L^2(\mu), g \in L^2(\nu)$ correspondingly. This will be exactly the same type of decomposition as in the "nonhomogeneous $T1$" theorems we met above. We use the notion of weighted Haar functions h_I^μ, h_I^ν, which we did not use above, and the notion of operators $\Delta_Q^\mu, \Delta_Q^\nu$, which has been already used above.

Let us recall that h_I^μ denotes the Haar function (supported by the interval $I \in \mathcal{D}^\mu$) with respect to measure μ. In other words, it has two values (one on the left half I_- of I, and one on the right half I_+ of I) such that

$$\int_I h_I^\mu \, d\mu = 0 \,,$$

$$\int_I (h_I^\mu)^2 \, d\mu = 1 \,.$$

The formula is

$$h_I^\mu = \frac{1}{\mu(I)^{1/2}} \Big[\Big(\frac{\mu(I_-)}{\mu(I_+)}\Big)^{1/2} \chi_{I_-} - \Big(\frac{\mu(I_-)}{\mu(I_+)}\Big)^{1/2} \chi_{I_+} \Big], \ I \in \mathcal{D}^\mu \,.$$

The same is true for h_I^ν with $\mathcal{D}^\nu$ replacing $\mathcal{D}^\mu$. We introduce the familiar operators $\Delta_I^\mu, \Delta_I^\nu$. Let $f \in L^2(\mu), g \in L^2(\nu)$ be two test functions as above:

$$\Delta_I^\mu(f) := (f, h_I^\mu)_\mu h_I^\mu, \ I \in \mathcal{D}^\mu, \ |I| \le 1 \,,$$

$$\Delta_I^\nu(g) := (g, h_I^\nu)_\nu h_I^\nu, \ I \in \mathcal{D}^\nu, \ |I| \le 1 \,.$$

Also, let I_0^μ denote the interval of $\mathcal{D}^\mu$ of length 1 containing supp(f), the same with I_0^ν changing f to g and μ to ν.

$$\Lambda^\mu(f) := (\int_{I_0^\mu} f \, d\mu) \, \chi_{I_0^\mu}, \ \Lambda^\nu(g) := (\int_{I_0^\nu} g \, d\nu) \, \chi_{I_0^\nu} \,.$$

It is easy to see that functions $\Lambda^\mu(f), \Delta_I^\mu(f), I \in \mathcal{D}^\mu$ are all pairwise orthogonal with respect to the scalar product $(\cdot, \cdot)_\mu$ of $L^2(\mu)$. The same is true for $\Lambda^\nu(f), \Delta_I^\nu(f), I \in \mathcal{D}^\nu$ with respect to the scalar product $(\cdot, \cdot)_\nu$ of $L^2(\nu)$. Actually, it is easy to see that the family $\chi_{I_0^\mu}, h_I^\mu, I \subset I_0^\mu$ is dense in the set of functions from $L^2(\mu)$ supported by $[\frac{1}{4}, \frac{3}{4}]$. The same is true if we replace μ by ν. Thus,

$$(17.1) \quad f = \Lambda^\mu(f) + \sum_{I \in \mathcal{D}^\mu, I \subset I_0^\mu} \Delta_I^\mu(f), \ \|f\|_\mu^2 = \|\Lambda^\mu(f)\|_\mu^2 + \sum_{I \in \mathcal{D}^\mu, I \subset I_0^\mu} \|\Delta_I^\mu(f)\|_\mu^2.$$

Similarly,

$$(17.2) \quad g = \Lambda^\nu(g) + \sum_{I \in \mathcal{D}^\mu, I \subset I_0^\mu} \Delta_I^\nu(g), \ \|g\|_\nu^2 = \|\Lambda^\nu(g)\|_\nu^2 + \sum_{I \in \mathcal{D}^\nu, I \subset I_0^\nu} \|\Delta_I^\nu(g)\|_\nu^2.$$

These decompositions and the assumptions (15.14),(15.15) imply in a very easy fashion that we can consider only the case

$$(17.3) \qquad \Lambda^\mu(f) = 0, \ \Lambda^\nu(g) = 0.$$

In fact, $(H_\mu f, g)_\nu = (H_\mu f - \Lambda^\mu(f), g)_\nu + (\int_{I_0^\mu} f \, d\mu)(H_\mu(\chi_{I_0^\mu}), g)_\nu$, and the second term is bounded by $C(C_\chi) \|f\|_\mu \|g\|_\nu$ trivially by (15.14). Using (15.15) one can get rid of $\Lambda^\nu(g)$ as well.

So we always work under the assumption (17.3). Now, for simplicity, we think that f, g are real valued. Then

$$(H_\mu f, g)_\nu = \sum_{I \in \mathcal{D}^\mu, J \in \mathcal{D}^\nu} (f, h_I^\mu)_\mu (H_\mu h_I^\mu, h_J^\nu)_\nu (g, h_I^\nu)_\nu.$$

17.1. Bad and good parts of f and g

We use "good-bad" decomposition of test functions f, g exactly as in Chapter 10.2. Consider two fixed lattices $\mathcal{D}^\mu, \mathcal{D}^\nu$ (so we fixed a point in Ω; see the notation above).

We call the interval $I \in \mathcal{D}^\mu$ bad if there exists $J \in \mathcal{D}^\nu$ such that

$$(17.4) \qquad |J| \geq |I|, \ \text{dist}(e(J), I) < |J|^{3/4}|I|^{1/4}.$$

Here $e(J) := \partial J \cup \text{mid point of } J$. Similarly, one defines bad intervals $J \in \mathcal{D}^\nu$.

Definition. We fix a large integer $r = C(C_\chi, C_d)$ to be chosen later, and we say that $I \in \mathcal{D}^\mu$ is *essentially bad* if there exists $J \in \mathcal{D}^\nu$ satisfying (17.4) such that it is much longer than I, namely, $|J| \geq 2^r |I|$.

If the interval is not essentially bad, it is called *good*.

Now

$$(17.5) \qquad f = f_{bad} + f_{good}, \ f_{bad} := \sum_{I \in \mathcal{D}^\mu, \, I \text{ is essentially bad}} \Delta_I^\mu f.$$

The same type of decomposition is used for g:

$$(17.6) \qquad g = g_{bad} + g_{good}, \ g_{bad} := \sum_{J \in \mathcal{D}^\nu, \, J \text{ is essentially bad}} \Delta_J^\nu g.$$

17.2. Estimates on good functions

We saw in Section 10.2 that it is enough to estimate $|(H_\mu f_{good}, g_{good})_\nu|$, because

$$(17.7) \qquad (H_\mu f, g)_\nu = (H_\mu f_{good}, g_{good})_\nu + (H_\mu f_{bad}, g_{good})_\nu + (H_\mu f, g_{bad})_\nu .$$

In Section 10.1 we formulated Theorem 10.3 (and in Section 13.1 it has been proved). Now we use the exact analog of this result. The proof is almost verbatim the same. In fact, we repeat the proof from Section 13.1, using the fact that

$$(17.8) \qquad \mathbb{P}\{(\omega_1, \omega_2) \in \Omega : I \text{ is essentially bad} \,|\, I \in \mathcal{D}^\mu\} \leq \tau(r) \to 0, \;\; r \to \infty .$$

This is an analog of Theorem 10.1 of Chapter 10.2. The proof follows from the proof of Theorem 10.1. The proof of the next result is even easier than the one in Section 13.1, because we do not need to use Lemma 12.5 here; our decomposition (17.1) is already orthogonal, and this simplifies the proof. So we consider the following result as already proved.

THEOREM 17.1. *We consider the decomposition of f to bad and good part, and take a bad part of it for every $\omega = (\omega_1, \omega_2) \in \Omega$. Let $\mathbb{E}$ denote the expectation with respect to $(\Omega, \mathbb{P})$. Then*

$$(17.9) \qquad \mathbb{E}(\|f_{bad}\|_\mu) \leq \varepsilon(r)\|f\|_\mu, \;\; where \;\; \varepsilon(r) \to 0, \;\; r \to \infty .$$

The same is true for g:

$$(17.10) \qquad \mathbb{E}(\|g_{bad}\|_\nu) \leq \varepsilon(r)\|g\|_\nu, \;\; where \;\; \varepsilon(r) \to 0, \;\; r \to \infty .$$

Coming back to (17.7), we get

$$|(H_\mu f, g)_\nu| \leq \mathbf{E}|(H_\mu f_{good}, g_{good})_\nu| + \|H_\mu\|\mathbf{E}\|f_{bad}\|_\mu\|g_{good}\|_\nu + \|H_\mu\|\|\mathbf{E}f\|_\mu\|g_{bad}\|_\nu$$
$$\leq \mathbf{E}|(H_\mu f_{good}, g_{good})_\nu| + 2C\varepsilon(r)\|f\|_\mu\|g\|_\nu,$$

where C denotes $\|H_\mu\|_{L^2(\mu)\to L^2(\nu)}$ (a priori finite; see Chapter 15). Choosing r to be such that $C\varepsilon(r) < \frac{1}{4}$, and choosing f, g to almost reach $C\|f\|_\mu\|g\|_\nu$ in $|(H_\mu f, g)_\nu|$, we get

$$\frac{1}{2}C\|f\|_\mu\|g\|_\nu \leq \mathbb{E}|(H_\mu f_{good}, g_{good})_\nu|$$

for these special f, g. If we manage to prove that *for all* f, g (see the notation for C_d, C_χ, C_p in Theorem 15.1)

$$(17.11) \quad |(H_\mu f_{good}, g_{good})_\nu| \leq C(C_d, C_\chi, C_p)\|f\|_\mu\|g\|_\nu \;\; \forall f \in L^2(\mu), \forall g \in L^2(\nu),$$

then we obtain

$$\|H_\mu\|_{L^2(\mu)\to L^2(\nu)} = C \leq 2C(C_d, C_\chi, C_p) ,$$

which finishes the proof of Theorem 15.1.

The rest is devoted to the proof of (17.11).

CHAPTER 18

Long Range Interaction

So let lattices $\mathcal{D}^\mu, \mathcal{D}^\nu$ be fixed, and let f, g be two good functions with respect to these lattices. Boundedness on characteristic functions declared in (15.14), (15.15) obviously imply

$$(18.1) \qquad |(H_\mu \Delta_I^\mu f, \Delta_J^\nu g)_\nu| \leq C_\chi \|\Delta_I^\mu f\|_\mu \|\Delta_J^\nu g\|_\nu \,.$$

Therefore, in the sum $(H_\mu f, g)_\nu = \sum_{I \in \mathcal{D}^\mu, J \in \mathcal{D}^\nu} (H_\mu \Delta_I^\mu, \Delta_J^\nu g)_\nu$, the "diagonal" part can be easily estimated. Namely, (below r is the number involved in the definition of good functions in the previous section, and we always have $I \in \mathcal{D}^\mu, J \in \mathcal{D}^\nu$ without mentioning this):

$$(18.2) \qquad \sum_{2^{-r}|J| \leq |I| \leq 2^r |J|, \mathrm{dist}(I,J) \leq \max(|I|,|J|)} |(H_\mu \Delta_I^\mu f, \Delta_J^\nu g)_\nu| \leq C(r, C_\chi) \|f\|_\mu \|g\|_\nu \,.$$

Let us consider the sums

$$(18.3) \quad \Sigma_1 := \sum_{2^{-r}|J| \leq |I| \leq |J|, \mathrm{dist}(I,J) \geq |J|} |(H_\mu \Delta_I^\mu f, \Delta_J^\nu g)_\nu| \leq C(r, C_\chi) \|f\|_\mu \|g\|_\nu \,.$$

$$(18.4) \quad \Sigma_2 := \sum_{2^{-r}|I| \leq |J| \leq |I|, \mathrm{dist}(I,J) \geq |I|} |(H_\mu \Delta_I^\mu f, \Delta_J^\nu g)_\nu| \leq C(r, C_\chi) \|f\|_\mu \|g\|_\nu \,.$$

They can be estimated in a symmetric fashion. So we will only deal with the first one.

LEMMA 18.1. *Let* $|I| \leq |J|$, $\mathrm{dist}(I,J) \geq |J|$. *Then*

$$(18.5) \qquad |(H_\mu \Delta_I^\mu f, \Delta_J^\nu g)_\nu| \leq A \frac{|I| \, \mu(I)^{1/2} \nu(J)^{1/2}}{(\mathrm{dist}(I,J) + |I| + |J|)^2} \|\Delta_I^\mu f\|_\mu \|\Delta_J^\nu g\|_\nu \,.$$

Proof. Let c be the center of I. We use the fact that $\int \Delta_I^\mu f \, d\mu = 0$ to write

$$(H_\mu \Delta_I^\mu f, \Delta_J^\nu g)_\nu = \int_I d\mu(t) \int_J d\nu(s) \frac{1}{t-s} \Delta_I^\mu f(t) \Delta_J^\nu g(s)$$

$$= \int_I d\mu(t) \int_J d\nu(s) \left(\frac{1}{t-s} - \frac{1}{c-s}\right) \Delta_I^\mu f(t) \Delta_J^\nu g(s) \,.$$

Then one can easily see that

$$(18.6) \qquad |(H_\mu \Delta_I^\mu f, \Delta_J^\nu g)_\nu| \leq A \int_I \int_J \frac{|I|}{|t-s|^2} |\Delta_I^\mu f(t)| |\Delta_J^\nu g(s)| d\mu(t) d\nu(s) \,.$$

Now we estimate the kernel $\frac{|I|}{|t-s|^2} \chi_I(t) \chi_J(s) \leq A \frac{|I|}{(\mathrm{dist}(I,J)+|I|+|J|)^2}$ using that $|I| \leq |J|$, $\mathrm{dist}(I,J) \geq |J|$. On the other hand,

$$\|\Delta_I^\mu f\|_{L^1(\mu)} \leq \mu(I)^{1/2} \|\Delta_I^\mu f\|_\mu, \ \|\Delta_J^\nu g\|_{L^1(\nu)} \leq \mu(J)^{1/2} \|\Delta_J^\nu g\|_\nu \,.$$

And the lemma is proved. $\qquad \square$

Let us notice that Lemma 18.1 allows us to write the following estimate for the sum of (18.3) (as usual $I \in \mathcal{D}^\mu, J \in \mathcal{D}^\nu$):

$$(18.7) \qquad \Sigma_1 \leq \sum_{n=0}^{\infty} 2^{-n} \sum_{I,J:|I|=2^{-n}|J|} \frac{|J|\,\mu(I)^{1/2}\nu(J)^{1/2}}{(\operatorname{dist}(I,J)+|I|+|J|)^2} \|\Delta_I^\mu f\|_\mu \|\Delta_J^\nu g\|_\nu \,.$$

Or Σ_1 is bounded by

$$(18.8) \qquad \sum_{n=0}^{\infty} 2^{-n} \sum_{k \in \mathbb{Z}} \sum_{I,J:|I|=2^{-n+k},|J|=2^k} \frac{2^k\,\mu(I)^{1/2}\nu(J)^{1/2}}{(\operatorname{dist}(I,J)+2^k)^2} \|\Delta_I^\mu f\|_\mu \|\Delta_J^\nu g\|_\nu \,.$$

To estimate "the n, k" slice

$$\Sigma_{n,k} := \sum_{I,J:|I|=2^{-n+k},|J|=2^k} \frac{2^k}{(\operatorname{dist}(I,J)+2^k)^2} \mu(I)^{1/2}\nu(J)^{1/2} \|\Delta_I^\mu f\|_\mu \|\Delta_J^\nu g\|_\nu \,,$$

let us introduce the following notation:

$$\varphi(t) = \sum_{I \in \mathcal{D}^\mu, |I|=2^{-n+k}} \frac{\|\Delta_I^\mu f\|_\mu}{\mu(I)^{1/2}} \chi_I(t), \quad \psi(s) = \sum_{J \in \mathcal{D}^\nu, |J|=2^k} \frac{\|\Delta_J^\nu g\|_\nu}{\nu(J)^{1/2}} \chi_J(s) \,.$$

Also

$$K_y(t,s) := \frac{y}{y^2 + |t-s|^2}, \quad y > 0, \ t,s \in \mathbb{R} \,.$$

Then

$$(18.9) \qquad \Sigma_{n,k} \leq \int_{\mathbb{R}} d\mu(t) \int_{\mathbb{R}} d\nu(s) K_{2^k}(t,s)\varphi(t)\psi(s) \,.$$

LEMMA 18.2. *The integral operator $f \to \int K_y(t,s)\varphi(t)\,d\mu(t)$ is bounded from $L^2(\mu)$ to $L^2(\nu)$ if $Q_{\mu,\nu}$ (recall that this quantity is equal to $\sup_{I \subset \mathbb{R}} \langle\mu\rangle_I \langle\nu\rangle_I$) is bounded. Its norm is bounded by $A\,Q_{\mu,\nu}^{1/2}$.*

Let us postpone the proof of this lemma, and let us finish the estimate of Σ_1 using it. First of all, the lemma gives the following estimate (notice that $Q_{\mu,\nu} \leq A\,C_p$) of $\Sigma_{n,k}$:

$$C(C_p)\|\varphi\|_\mu\|\psi\|_\nu = C(C_p)\Big(\sum_{I \in \mathcal{D}^\mu, |I|=2^{-n+k}} \|\Delta_I^\mu f\|_\mu^2\Big)^{1/2} \Big(\sum_{J \in \mathcal{D}^\nu, |J|=2^k} \|\Delta_J^\nu g\|_\nu^2\Big)^{1/2} \,.$$

By Cauchy inequality,

$$\sum_k \Sigma_{n,k} \leq \sum_k \Big(\sum_{J \in \mathcal{D}^\nu, |J|=2^k} \|\Delta_J^\nu g\|_\nu^2\Big)^{1/2} \Big(\sum_{I \in \mathcal{D}^\mu, |I|=2^{-n+k}} \|\Delta_I^\mu f\|_\mu^2\Big)^{1/2}$$

$$\leq \Big(\sum_{J \in \mathcal{D}^\nu} \|\Delta_J^\nu g\|_\nu^2\Big)^{1/2} \Big(\sum_{I \in \mathcal{D}^\mu} \|\Delta_I^\mu f\|_\mu^2\Big)^{1/2} \leq \|f\|_\mu \|g\|_\nu$$

by (17.1). Then (18.8) gives $\Sigma_1 \leq \sum_{n=0}^{\infty} 2^{-k} \sum_k \Sigma_{n,k}$, and so

$$\Sigma_1 \leq C(C_p) \sum_{n=0}^{\infty} 2^{-n} \|f\|_\mu \|g\|_\nu = 2C(C_p)\|f\|_\mu\|g\|_\nu,$$

and our long range interaction sum Σ_1 is finally estimated.

Proof of Lemma 18.2. Let us consider several other averaging operators. One of them is

$$I\varphi(s) := \int \chi_{[-\frac{1}{2}, \frac{1}{2}]}(s-t)\varphi(t)\, d\mu(t)\,.$$

Another is as follows: let G be all intervals ℓ_k of the type $[2k, 2k+2]$, $k \in \mathbb{Z}$. Consider

$$A_G\varphi(s) := \sum_k \chi_{\ell_k}(s)\frac{1}{|\ell_k|}\int_{\ell_k}\varphi\, d\mu\,.$$

Consider also shifted grid $G(x) = G + x$, $x \in [0,2)$, and corresponding $A_{G(x)}$.

Notice that

$$(18.10) \qquad\qquad I\varphi(s) \leq a\int_0^2 A_{G(x)}\varphi(s)\, dx\,.$$

In fact, consider $[0,2]$, $\frac{1}{2}dx$ as an obvious probability space of all grids $G(x)$. Then it is easy to see that for every s the unit interval $[s - \frac{1}{2}, s + \frac{1}{2}]$ with probability at least $1/2$ is a subinterval of one of the intervals of $G(x)$. Then the above inequality becomes obvious (and $a = 4$).

On the other hand, the norm of operator A_G as an operator from $L^2(\mu)$ to $L^2(\nu)$ is bounded by $2Q_{\mu,\nu}^{1/2}$. In fact, if $\ell_k = [2k, 2k+2]$, then

$$\|A_G\varphi\|_\nu^2 \leq \sum_k \Big(\int_{\ell_k}|\varphi|\, d\mu\Big)^2\nu(\ell_k) \leq \sum_k \Big(\int_{\ell_k}|\varphi|^2\, d\mu\Big)\nu(\ell_k)\mu(\ell_k)$$

$$\leq 4Q_{\mu,\nu}\sum_k\int_{\ell_k}|\varphi|^2\, d\mu = 4Q_{\mu,\nu}\|f\|_\mu^2\,.$$

The same, of course, can be said about $\|A_{G(x)}\varphi\|_\nu^2$ for any x. Then (18.10) implies that the norm of averaging operator I from $L^2(\mu)$ to $L^2(\nu)$ is bounded by $AQ_{\mu,\nu}^{1/2}$. Let us call by I_r the operator of the same type as I, but the convolution now will be with the normalized characteristic function of the interval $[-r, r]$:

$$I_r\varphi(s) := \frac{1}{2r}\int \chi_{[-r,r]}(s-t)\varphi(t)\, d\mu(t)\,.$$

It is obvious that the reasoning above can be repeated without any change and we get

$$(18.11) \qquad\qquad \|I_r\varphi\|_\nu^2 \leq A\, Q_{\mu,\nu}\|f\|_\mu^2\,.$$

To finish with the operator given by $f \to \int K_y(t,s)\varphi(t)\, d\mu(t)$ as an operator from $L^2(\mu)$ to $L^2(\nu)$, let us notice that (and this is a standard inequality for the Poisson kernel)

$$\int K_y(t,s)|\varphi(t)|\, d\mu(t) \leq A\sum_{k=0}^\infty 2^{-k}(I_{y\cdot 2^k}|\varphi|)(s)\,.$$

Now Lemma 18.2 follows immediately from (18.11) and the last inequality.

CHAPTER 19

The Rest of the Long Range Interaction

As always, all I's below are in $\mathcal{D}^{\mu}$; all J's below are in $\mathcal{D}^{\nu}$. Consider now the following two sums:

$$(19.1) \qquad \sigma_1 := \sum_{|I| < 2^{-r}|J|, I \cap J = \emptyset} |(H_{\mu}\Delta_I^{\mu}f, \Delta_J^{\nu}g)_{\nu}|,$$

$$(19.2) \qquad \sigma_2 := \sum_{|J| < 2^{-r}|I|, I \cap J = \emptyset} |(H_{\mu}\Delta_I^{\mu}f, \Delta_J^{\nu}g)_{\nu}|.$$

They can be estimated in a symmetric fashion. So we will only deal with the first one.

Notice that f, g are good functions. These means, in particular, that I, J, which we meet in (19.1), satisfy

$$(19.3) \qquad \mathrm{dist}(I, \partial J) \geq |J|^{3/4}|I|^{1/4}.$$

This is just (17.4) for disjoint I, J with I *not* essentially bad (see the definition at the beginning of Section 17.1).

LEMMA 19.1. *Let I, J be disjoint, $|I| < 2^{-r}|J|$, and satisfy (19.3). Then*

$$(19.4) \qquad |(H_{\mu}\Delta_I^{\mu}f, \Delta_J^{\nu}g)_{\nu}| \leq A \frac{|I|^{1/2}|J|^{1/2}\,\mu(I)^{1/2}\nu(J)^{1/2}}{(\mathrm{dist}(I,J) + |I| + |J|)^2}\|\Delta_I^{\mu}f\|_{\mu}\|\Delta_J^{\nu}g\|_{\nu}.$$

Proof. If $\mathrm{dist}(I, J) \geq |J|$, this has been already proved in Lemma 18.1. So let $\mathrm{dist}(I, J) \leq |J|$, I, J being disjoint. Repeating (18.6), one gets

$$|(H_{\mu}\Delta_I^{\mu}f, \Delta_J^{\nu}g)_{\nu}| \leq A \int_I \int_J \frac{|I|}{|t-s|^2}|\Delta_I^{\mu}f(t)||\Delta_J^{\nu}g(s)|d\mu(t)d\nu(s).$$

Now we estimate the kernel $\frac{|I|}{|t-s|^2}\chi_I(t)\chi_J(s) \leq A \frac{|I|}{\mathrm{dist}(I,\partial J)^2}$. Therefore,

$$(19.5) \qquad |(H_{\mu}\Delta_I^{\mu}f, \Delta_J^{\nu}g)_{\nu}| \leq A \frac{|I|}{\mathrm{dist}(I,\partial J)^2}\mu(I)^{1/2}\nu(J)^{1/2}\|\Delta_I^{\mu}f\|_{\mu}\|\Delta_J^{\nu}g\|_{\nu}.$$

We use (19.3) to write

$$\frac{|I|}{\mathrm{dist}(I,\partial J)^2} \leq \frac{|I|^{1/2}}{|J|^{3/2}} = \frac{|I|^{1/2}|J|^{1/2}}{|J|^2} \leq A \frac{|I|^{1/2}|J|^{1/2}}{(\mathrm{dist}(I,J) + |I| + |J|)^2},$$

because we assumed $\mathrm{dist}(I, J) \leq |J|$ and I is shorter than J. This inequality and (19.5) finish the proof of the lemma. $\qquad \square$

Let us notice that Lemma 19.1 allows us to write the following estimate for the sum σ_1 from (19.1):

(19.6)
$$\sigma_1 \leq \sum_{n=0}^{\infty} 2^{-n/2} \sum_{I,J:|I|=2^{-n}|J|} \frac{|J|}{(\operatorname{dist}(I,J)+|I|+|J|)^2} \mu(I)^{1/2}\nu(J)^{1/2}\|\Delta_I^\mu f\|_\mu \|\Delta_J^\nu g\|_\nu \,.$$

Or σ_1 is bounded by

(19.7)
$$\sum_{n=0}^{\infty} 2^{-n/2} \sum_{k\in\mathbb{Z}} \sum_{I,J:|I|=2^{-n+k},|J|=2^k} \frac{2^k}{(\operatorname{dist}(I,J)+2^k)^2}\mu(I)^{1/2}\nu(J)^{1/2}\|\Delta_I^\mu f\|_\mu \|\Delta_J^\nu g\|_\nu \,.$$

To estimate "the n,k" slice

$$\sigma_{n,k} := \sum_{I,J:|I|=2^{-n+k},|J|=2^k} \frac{2^k}{(\operatorname{dist}(I,J)+2^k)^2}\mu(I)^{1/2}\nu(J)^{1/2}\|\Delta_I^\mu f\|_\mu \|\Delta_J^\nu g\|_\nu \,,$$

let us use again the notation

$$\varphi(t) = \sum_{I\in\mathcal{D}^\mu,|I|=2^{-n+k}} \frac{\|\Delta_I^\mu f\|_\mu}{\mu(I)^{1/2}}\chi_I(t), \; \psi(s) = \sum_{J\in\mathcal{D}^\nu,|I|=2^k} \frac{\|\Delta_J^\nu g\|_\nu}{\nu(J)^{1/2}}\chi_J(s)\,.$$

Also

$$K_y(t,s) := \frac{y}{y^2+|t-s|^2}, \; y>0, \; t,s\in\mathbb{R}\,.$$

Then

(19.8)
$$\sigma_{n,k} \leq \int_{\mathbb{R}} d\mu(t) \int_{\mathbb{R}} d\nu(s) K_{2^k}(t,s)\varphi(t)\psi(s)\,.$$

Lemma 18.2 now gives as before the estimate of σ_1. First of all, the lemma gives the following estimate (notice that $Q_{\mu,\nu} \leq A\,C_p$) of $\sigma_{n,k}$:

$$C\|\varphi\|_\mu\|\psi\|_\nu = C(C_p)\Big(\sum_{I\in\mathcal{D}^\mu,|I|=2^{-n+k}} \|\Delta_I^\mu f\|_\mu^2\Big)^{1/2}\Big(\sum_{J\in\mathcal{D}^\nu,|J|=2^k} \|\Delta_J^\nu g\|_\nu^2\Big)^{1/2}\,.$$

By Cauchy inequality,

$$\sum_k \sigma_{n,k} \leq \sum_k \Big(\sum_{J\in\mathcal{D}^\nu,|J|=2^k} \|\Delta_J^\nu g\|_\nu^2\Big)^{1/2}\Big(\sum_{I\in\mathcal{D}^\mu,|I|=2^{-n+k}} \|\Delta_I^\mu f\|_\mu^2\Big)^{1/2}$$

$$\leq \Big(\sum_{J\in\mathcal{D}^\nu} \|\Delta_J^\nu g\|_\nu^2\Big)^{1/2}\Big(\sum_{I\in\mathcal{D}^\mu} \|\Delta_I^\mu f\|_\mu^2\Big)^{1/2} \leq \|f\|_\mu\|g\|_\nu$$

by (17.1). Then (19.7) gives $\sigma_1 \leq \sum_{n=0}^{\infty} 2^{-n/2}\sum_k \sigma_{n,k}$, and so

$$\sigma_1 \leq C(C_p)\sum_{n=0}^{\infty} 2^{-n/2}\|f\|_\mu\|g\|_\nu = A\,C(C_p)\,\|f\|_\mu\|g\|_\nu,$$

and our long range interaction sum σ_1 is finally estimated. Symmetric estimate holds for σ_2 from (19.2).

Conclusion: If f,g are good, then the sum of all terms $|(H_\mu\Delta_I^\mu f, \Delta_J^\nu g)_\nu|$ such that either $\frac{|I|}{|J|} \in [2^{-r}, 2^r]$ or $I\cap J = \emptyset$ has the correct estimate $C(C_p)\|f\|_\mu\|g\|_\nu$.

CHAPTER 20

The Short Range Interaction

As always, all I's below are in $\mathcal{D}^\mu$; all J's below are in $\mathcal{D}^\nu$.

Let us consider the sums

$$(20.1) \qquad \rho := \sum_{|I|<2^{-r}|J|,I\subset J,\mathrm{dist}(I,e(J))\geq|J|^{3/4}|I|^{1/4}} (\Delta_I^\mu f, H_\nu \Delta_J^\nu g)_\mu\,,$$

$$(20.2) \qquad \tau := \sum_{|J|<2^{-r}|I|,J\subset I,\mathrm{dist}(J,e(I))\geq|I|^{3/4}|J|^{1/4}} (H_\mu \Delta_I^\mu f, \Delta_J^\nu g)_\nu\,.$$

They can be estimated in a symmetric fashion. So we will only deal with, say, the second one. It is very important that unlike the sums Σ_i, σ_i, this sum does not have absolute value on *each* term.

Consider each term of τ and split it into three terms. To do this, let I_i denote the half of I, which contains J. And I_n is another half. Let $\hat{I}$ denote an arbitrary superinterval of I in the same lattice: $\hat{I} \in \mathcal{D}^\mu$.

We write

$$(H_\mu \Delta_I^\mu f, \Delta_J^\nu g)_\nu = (H_\mu(\chi_{I_n}\Delta_I^\mu f), \Delta_J^\nu g)_\nu + (H_\mu(\chi_{I_i}\Delta_I^\mu f), \Delta_J^\nu g)_\nu$$

$$(20.3) \qquad = (H_\mu(\chi_{I_n}\Delta_I^\mu f), \Delta_J^\nu g)_\nu + \langle\Delta_I^\mu f\rangle_{\mu,I_i}(H_\mu(\chi_{\hat{I}}), \Delta_J^\nu g)_\nu$$

$$- \langle\Delta_I^\mu f\rangle_{\mu,I_i}(H_\mu(\chi_{\hat{I}\setminus I_i}), \Delta_J^\nu g)_\nu\,.$$

Here $\langle\Delta_I^\mu f\rangle_{\mu,I_i}$ is the average of $\Delta_I^\mu f$ with respect to μ over I_i, which is the same value of this function on I_i (by construction $\Delta_I^\mu f$ assumes on I two values, one on I_i, one on I_n).

Definition. We call them as follows: the first one is "the neighbor-term", the second one is "the difficult term", and the third one is "the stopping term".

20.1. The estimate of neighbor-terms

We have the same estimate as in Lemma 19.1:

$$|(H_\mu(\chi_{I_n}\Delta_I^\mu f), \Delta_J^\nu g)_\nu|$$

$$(20.4) \qquad \leq \frac{A\,|I|^{1/2}|J|^{1/2}}{(\mathrm{dist}(I,J)+|I|+|J|)^2}\mu(I)^{1/2}\nu(J)^{1/2}\|\Delta_I^\mu f\chi_{I_n}\|_\mu\|\Delta_J^\nu g\|_\nu\,.$$

Of course, $\|\chi_{I_n}\Delta_I^\mu f\|_\mu \leq \|\Delta_I^\mu f\|_\mu$. So the estimate of the sum of absolute values of neighbor-terms is exactly the same as the estimate of σ_1 in the preceding section.

20.2. The estimate of stopping terms

Here the fact that we deal with the Hilbert transform is used in an essential way. The estimate for other Calderón–Zygmund kernels will require some new tricks. We need the following definition.

Definition. Given an interval $I = [a, b]$ and any measure $d\sigma$ on the real line, we write

$$P_{[a,b]}d\sigma := \frac{1}{\pi} \int_{\mathbb{R}} \frac{b-a}{(b-a)^2 + ((b+a)/2 - t)^2} \, d\sigma(t) \,.$$

This is the Poisson integral at the point where the real part is the center of the interval, and the imaginary part is the length of the interval.

We want to estimate

$$|\langle \Delta_I^\mu f\rangle_{\mu, I_i}||(H_\mu(\chi_{\hat{I}\setminus I}), \Delta_J^\nu g)_\nu| \,.$$

First of all,

$$|\langle \Delta_I^\mu f\rangle_{\mu, I_i}| \le \frac{\|\Delta_I^\mu f\|_\mu}{\mu(I_i)^{1/2}} \,.$$

Secondly,

$$|(H_\mu(\chi_{\hat{I}\setminus I}), \Delta_J^\nu g)_\nu| = |(\chi_{\hat{I}\setminus I}, H_\nu \Delta_J^\nu g)_\mu|$$

$$\le A \left(\int_{\hat{I}\setminus I} d\mu(x) \frac{|J|}{\operatorname{dist}(x, J)^2} \right) \|\Delta_J^\nu g\|_{L^1(\nu)} \,.$$

This is the usual trick with subtraction of the kernel; it uses the fact that we have the following cancellation:

$$\int \Delta_J^\nu g \, d\nu = 0 \,.$$

We continue by denoting the center of I_i by c:

$$\le A\, \nu(J)^{1/2}\|\Delta_J^\nu g\|_\nu \int_{\hat{I}\setminus I} \frac{\operatorname{dist}(x, c)^2}{\operatorname{dist}(x, J)^2} \frac{|J|}{\operatorname{dist}(x, c)^2} \, d\mu(x)$$

$$\le A\, \nu(J)^{1/2}\|\Delta_J^\nu g\|_\nu \int_{\hat{I}\setminus I} \frac{\operatorname{dist}(e(I), c)^2}{\operatorname{dist}(e(I), J)^2} \frac{|J|}{\operatorname{dist}(x, c)^2} \, d\mu(x) \,,$$

where $e(I)$ is the set consisting of two ends and the center of I. The elementary inequality above uses of course the specific nature of the Hilbert transform. We continue, using the definition above,

$$\le A\, \nu(J)^{1/2}\|\Delta_J^\nu g\|_\nu \int_{\hat{I}\setminus I} \frac{|I|^2|J|}{|I|^{3/2}|J|^{1/2}} \frac{1}{\operatorname{dist}(x, c)^2} \, d\mu(x)$$

$$\le A\, \nu(J)^{1/2}\|\Delta_J^\nu g\|_\nu \left(\frac{|J|}{|I|}\right)^{1/2} P_{I_i}(\chi_{\hat{I}\setminus I}\, d\mu) \,.$$

Thus

(20.5) $$\qquad |(H_\mu(\chi_{\hat{I}\setminus I}), \Delta_J^\nu g)_\nu| \le A\, \nu(J)^{1/2}\|\Delta_J^\nu g\|_\nu \left(\frac{|J|}{|I|}\right)^{1/2} P_{I_i}(\chi_{\hat{I}\setminus I}\, d\mu) \,.$$

We now get the estimate of the stopping term:

$$|\langle \Delta_I^\mu f\rangle_{\mu, I_i}||(H_\mu(\chi_{\hat{I}\setminus I}), \Delta_J^\nu g)_\nu|$$

(20.6) $$\qquad \le A \left(\frac{\nu(J)}{\mu(I_i)}\right)^{1/2} \left(\frac{|J|}{|I|}\right)^{1/2} P_{I_i}(\chi_{\hat{I}\setminus I_i}\, d\mu)\|\Delta_J^\nu g\|_\nu\|\Delta_I^\mu f\|_\mu \,.$$

20.3. The choice of stopping intervals

Let K be a large constant, $K = K(C_\chi, C_d, C_p)$ to be chosen later. Fix an interval $\hat{I} \in \mathcal{D}^\mu$. Let us call its *subinterval* $I \in \mathcal{D}^\mu$ a *stopping interval* if it is the last one (by going to the smaller ones by inclusion) such that for both its sons I_1, I_2,

$$(20.7) \qquad \left[P_{I_i}(\chi_{\hat{I} \setminus I_i} \, d\mu) \right]^2 \nu(I_i) \le K \, \mu(I_i), \ i = 1, 2.$$

Here is the place where we use the doubling measure properties:

THEOREM 20.1. *If μ, ν have doubling property and supports coinciding with $\mathbb{R}$, then for every $\hat{I} \in \mathcal{D}^\mu$,*

$$(20.8) \qquad \sum_{I \in \mathcal{D}^\mu, \, I \subset \hat{I}, \, I \text{ is maximal stopping}} \mu(I) \le \frac{1}{2}\mu(\hat{I}),$$

provided that the constant K in the stopping criterion (20.7) is large enough.

Proof. Notation. Let interval S be in $\mathcal{D}^\mu$. We denote by Q_S the square built on S as a base, which lies in the upper half plane $\mathbb{C}_+$. The symbol $\widehat{\mathbb{P}}_{\nu, Q_S}$ denotes the orthogonal projection in $L^2(\nu)$ onto functions supported by S, and having zero ν average over S.

LEMMA 20.2. *If ν has doubling property and support coinciding with $\mathbb{R}$, then for any $S \subset \hat{S}; S, \hat{S} \in \mathcal{D}^\mu$ one has*

$$(20.9) \qquad \left[P_S(\chi_{\hat{S} \setminus S} \, d\mu) \right]^2 \nu(S) \le C_d \| \widehat{\mathbb{P}}_{\nu, Q_S}(H_\mu \chi_{\hat{S} \setminus S}) \|_\nu^2.$$

Proof. We have

$$\| \widehat{\mathbb{P}}_{\nu, Q_S}(H_\mu \chi_{\hat{S} \setminus S}) \|_\nu^2 = \left| \sup_{\|\varphi\|_\nu \le 1, \int \varphi \, d\nu = 0, \text{supp}\, \varphi \subset S} (\varphi, H_\mu \chi_{\hat{S} \setminus S})_\nu \right|^2.$$

Take the following special φ—here the doubling will be used! We need to use the notation. Let $S := [A, E]$. Split it into three equal parts by C and D, $|AC| = |CD| = |DE|$. Let us assume, for example, that

$$\nu(DE) \le \nu(AC).$$

The doubling property of ν implies that there exists $\gamma = \gamma(C_d) > 0$ such that

$$\nu(DE) = \gamma \nu(AE) = \gamma \nu(S).$$

Of course, the "smoothness" of ν implies $\gamma \le \frac{1}{2} - \tau(C_d), \tau > 0$. In fact, CD has measure comparable with $\nu(S)$. So $\nu(DE) \le \frac{1}{2}(\nu(S) - \nu(CD))$. Now let $B \in AC$ be such that

$$\nu(AB) = \gamma \nu(S).$$

Let the function ψ be defined as follows:

$$\psi = 1 \text{ on } AB, \ \psi = 0 \text{ on } BD, \ \psi = -1 \text{ on } DE.$$

Then $\text{supp}\,\psi \subset S$, $\int \psi \, d\nu = 0$, $\int \psi^2 \, d\nu = 2\gamma \nu(S)$. Put

$$\varphi := \frac{\psi}{(2\gamma \nu(S))^{1/2}}.$$

Then

$$(20.10) \quad \|\widehat{\mathbb{P}}_{\nu,Q_s}(H_\mu\chi_{\hat{S}\setminus S})\|_\nu^2 \geq \left|(\varphi, H_\mu\chi_{\hat{S}\setminus S})_\nu\right|^2 = \frac{\gamma\nu(S)}{2}\left|(\frac{\psi}{\gamma\nu(S)}, H_\mu\chi_{\hat{S}\setminus S})_\nu\right|^2.$$

Let us denote by L the left part of $\hat{S}\setminus s$, and by R its right part. Then the quantity above can be estimated as follows:

$$\left(\frac{\psi}{\gamma\nu(S)}, H_\mu\chi_{\hat{S}\setminus S}\right)_\nu = \left[\int_L d\mu(x)\int_{AB}\frac{d\nu(y)/\gamma\nu(S)}{y-x} - \int_L d\mu(x)\int_{DE}\frac{d\nu(y)/\gamma\nu(S)}{y-x}\right]$$

$$+ \left[\int_R d\mu(x)\int_{AB}\frac{d\nu(y)/\gamma\nu(S)}{y-x} - \int_R d\mu(x)\int_{DE}\frac{d\nu(y)/\gamma\nu(S)}{y-x}\right]$$

$$= \left[\int_L d\mu(x)\frac{1}{y_{L,AB}-x} - \int_L d\mu(x)\frac{1}{y_{L,DE}-x}\right]$$

$$+ \left[\int_R d\mu(x)\frac{1}{y_{R,AB}-x} - \int_R d\mu(x)\frac{1}{y_{R,DE}-x}\right],$$

where $y_{L,AB}, y_{R,AB} \in AB$; $y_{L,DE}, y_{R,DE} \in DE$. We just used the continuity of $y \to \frac{1}{y-x}$, $y \in AB$ or $y \in DE$, $x \in R$ or $x \in L$. Points $y_{L,AB}, y_{R,AB}, y_{L,DE}, y_{R,DE}$ of course depend on x. Now for all $x \in L \cup R$,

$$y_{L,DE} - y_{L,AB} \geq |BD| \geq \frac{1}{3}|S|, \ y_{R,DE} - y_{R,AB} \geq |BD| \geq \frac{1}{3}|S|.$$

This is why we can continue our estimate as follows (here C is the same point $C \in S$ as above)

$$\left(\frac{\psi}{\gamma\nu(S)}, H_\mu\chi_{\hat{S}\setminus S}\right)_\nu \geq A\int_L\frac{|S|}{|C-x|^2}\,d\mu(x) + A\int_R\frac{|S|}{|C-x|^2}\,d\mu(x) \geq A'P_S(\chi_{\hat{S}\setminus S}d\mu).$$

Combining this inequality with (20.10), we get

$$(20.11) \quad \|\widehat{\mathbb{P}}_{\nu,Q_s}(H_\mu\chi_{\hat{S}\setminus S})\|_\nu^2 \geq \gamma A'\,(P_S(\chi_{\hat{S}\setminus S}d\mu))^2\nu(S).$$

This proves Lemma 20.2. □

We can continue the proof of the theorem. Let $\hat{I}$ be given, and let $\{I^k\}_{k=1}^M$ be maximal stopping subintervals of it. They are disjoint. In each of them there exists S^k, the grandson of I^k, such that

$$(P_{S^k}(\chi_{\hat{I}\setminus S^k}d\mu))^2\nu(S^k) > K\,\mu(S^k).$$

By Lemma 20.2 we can write

$$K\,\mu(S^k) \leq C(C_d)\,\|\widehat{\mathbb{P}}_{\nu,Q_{S^k}}(H_\mu\chi_{\hat{I}\setminus S^k})\|_\nu^2.$$

Let us sum this up. In what follows we use the orthogonality of $\widehat{\mathbb{P}}_{\nu,Q_{S^k}}$. We also use our *test functions assumptions* (15.14) from the main Theorem 15.1:

$$K\,\mu(S^k) \leq C(C_d)\sum_k\|\ldots\|_\nu^2 \leq 2C(C_d)\sum_k\|\widehat{\mathbb{P}}_{\nu,Q_{S^k}}(H_\mu\chi_{\hat{I}})\|_\nu^2$$

$$+ 2C(C_d)\sum_k\|\widehat{\mathbb{P}}_{\nu,Q_{S^k}}(H_\mu\chi_{S^k})\|_\nu^2$$

$$\leq 2C(C_d)\,\|(\sum_k\widehat{\mathbb{P}}_{\nu,Q_{S^k}})(H_\mu\chi_{\hat{I}})\|_\nu^2 + 2C(C_d)\sum_k\|H_\mu\chi_{S^k}\|_\nu^2$$

$$\leq 2C(C_d, C_\chi)\mu(\hat{I}) + 2C(C_d, C_\chi)\sum_k\mu(S^k).$$

If K is chosen to be (B is a large number)

$$K \geq (B+2)\,C(C_d, C_\chi)\,,$$

, then the last inequality implies

$$\sum_k \mu(S^k) \leq \frac{1}{B}\mu(\hat{I})\,.$$

But S^k are grandsons of I^k, and we can again use doubling—but this time for measure μ—to conclude

$$\sum_k \mu(I^k) \leq \frac{1}{2}\mu(\hat{I})\,,$$

if B is large enough. This is inequality (20.8). So, Theorem 20.1 is completely proved. $\qquad\square$

In Chapter 20 we introduced the sum, which we are left to estimate:

$$(20.12) \qquad \tau := \sum_{|J| < 2^{-r}|I|,\, J \subset I,\, \mathrm{dist}(J, e(I)) \geq |I|^{3/4}|J|^{1/4}} (H_\mu \Delta_I^\mu f, \Delta_J^\nu g)_\nu \,.$$

Each term of τ was decomposed into three terms. We recall: let I_i denote the half of I that contains J. And I_n is another half. Let $\hat{I}$ denote an arbitrary superinterval of I in the same lattice: $\hat{I} \in \mathcal{D}^\mu$.

We write

$$\begin{aligned}
(H_\mu \Delta_I^\mu f, \Delta_J^\nu g)_\nu &= (H_\mu(\chi_{I_n} \Delta_I^\mu f), \Delta_J^\nu g)_\nu + (H_\mu(\chi_{I_i} \Delta_I^\mu f), \Delta_J^\nu g)_\nu \\
&= (H_\mu(\chi_{I_n} \Delta_I^\mu f), \Delta_J^\nu g)_\nu + \langle \Delta_I^\mu f \rangle_{\mu, I_i}(H_\mu(\chi_{\hat{I}}), \Delta_J^\nu g)_\nu \\
&\quad - \langle \Delta_I^\mu f \rangle_{\mu, I_i}(H_\mu(\chi_{\hat{I} \setminus I_i}), \Delta_J^\nu g)_\nu \,.
\end{aligned}$$

Here $\langle \Delta_I^\mu f \rangle_{\mu, I_i}$ is the average of $\Delta_I^\mu f$ with respect to μ over I_i, which is the same as the value of this function on I_i (by construction $\Delta_I^\mu f$ assumes on I two values, one on I_i, one on I_n).

We call them as follows: the first one is "the neighbor-term", the second one is "the difficult term", and the third one is "the stopping term".

In what follows it is convenient to think that we consider our problem on the circle $\mathbb{T}$ rather than on the line. We want to explain how to choose $\hat{I}$ in stopping terms above.

We choose first $\hat{I} = \mathbb{T}$ (this is why the circle is more convenient; we have the first "hat" interval). We choose its maximal stopping subintervals $\{I\}$. Just use the criterion (20.7) from Section 20.3. Call each of these I's by the name $\hat{S}$. In each $\hat{S}$ again find its maximal stopping subintervals $\{S\}$. Et cetera .... All intervals, which were thus built, we call "stopping intervals". They have their generation. Stopping intervals, as a rule, will be denoted by symbols with "hats".

To explain the choice of $\hat{I}$ in stopping terms above, we need the notation.

Notation. If $\hat{S} \in \mathcal{D}^\mu$ is a stopping interval, and $\mathcal{S} = \{S\}, S \in \mathcal{D}^\mu$ is a collection of its maximal stopping subintervals, we call $\mathcal{O}_{\hat{S}}$ the collection of all intervals I from *both* lattices $\mathcal{D}^\mu, \mathcal{D}^\nu$, such that the top side of the square Q_I lies in the set $\Omega_{\hat{S}} := (\bar{Q}_{\hat{S}} \setminus \cup_{S \in \mathcal{S}} \bar{Q}_S)$.

The choice of $\hat{I}$ in a stopping term above is as follows: we choose the first (and unique) stopping interval $\hat{S}$ such that $I \in \Omega_{\hat{S}}$. We just put $\hat{I} = \hat{S}$.

Let us introduce the sum of absolute values of the "stopping terms" of the sum τ above (as always $I \in \mathcal{D}^\mu, J \in \mathcal{D}^\nu$):

$$t := \sum_{|J|<2^{-r}|I|, J \subset I, \mathrm{dist}(J,e(I)) \geq |I|^{3/4}|J|^{1/4}} |\langle \Delta_I^\mu f \rangle_{\mu, I_i}| |(H_\mu(\chi_{\hat{I} \setminus I_i}), \Delta_J^\nu g)_\nu|.$$

To estimate it we can use (20.6). Then (recall that I_i is the half of I containing J)

$$t \leq AT, \quad T := \sum_{|J|<2^{-r}|I|, J \subset I, \mathrm{dist}(J,e(I)) \geq |I|^{3/4}|J|^{1/4}} \left(\frac{\nu(J)}{\mu(I_i)}\right)^{1/2} \left(\frac{|J|}{|I|}\right)^{1/2}$$
$$\times P_{I_i}(\chi_{\hat{I} \setminus I_i} d\mu) \|\Delta_J^\nu g\|_\nu \|\Delta_I^\mu f\|_\mu.$$

We will follow the steps of [43] (and we will use the stopping criterion (20.7) based on constant K) to prove the following theorem.

THEOREM 20.3.
$$T \leq C(K) \|f\|_\mu \|g\|_\nu.$$

Proof. Put

$$r_{n,k} := \sum_{|J|<2^{-r}|I|, J \subset I_i, |I|=2^k, |J|=2^{-n+k}} \left(\frac{\nu(J)}{\mu(I_i)}\right)^{1/2} P_{I_i}(\chi_{\hat{I} \setminus I_i} d\mu) \|\Delta_J^\nu g\|_\nu \|\Delta_I^\mu f\|_\mu.$$

Then abusing slightly the notation we denote the halves of I by I_1, I_2. We get

$$r_{n,k} \leq \sum_{i=1}^2 \sum_{|I|=2^k} \|\Delta_I^\mu f\|_\mu \sum_{J \subset I_i, |J|=2^{-n+k}} \left(\frac{\nu(J)}{\mu(I_i)}\right)^{1/2} P_{I_i}(\chi_{\hat{I} \setminus I_i} d\mu) \|\Delta_J^\nu g\|_\nu.$$

Consider only I_1. By the Cauchy inequality, the estimate will be

$$\sum_{|I|=2^k} \|\Delta_I^\mu f\|_\mu \left(\sum_{J \subset I_1, |J|=2^{-n+k}} \left(\frac{\nu(J)}{\mu(I_1)}\right) [P_{I_1}(\chi_{\hat{I} \setminus I_1} d\mu)]^2 \right)^{1/2}$$
$$\times \left(\sum_{J \subset I_1, |J|=2^{-n+k}} \|\Delta_J^\nu g\|_\nu^2 \right)^{1/2}.$$

The middle term is bounded by $[P_{I_1}(\chi_{\hat{I} \setminus I_1} d\mu)]^2 \nu(I_1)/\mu(I_1)$. By (20.7) we get that the middle term is bounded by K. In fact, this was our choice of $\hat{I}$, which ensures that $I \in \mathcal{O}_{\hat{I}}$, and so (20.7) holds.

Thus, the last expression above is bounded by (this is just the Cauchy inequality)

$$K \sum_{|I|=2^k} \|\Delta_I^\mu f\|_\mu \left(\sum_{J \subset I_1, |J|=2^{-n+k}} \|\Delta_J^\nu g\|_\nu^2 \right)^{1/2}$$
$$\leq K \left(\sum_{|I|=2^k} \|\Delta_I^\mu f\|_\mu^2 \right)^{1/2} \left(\sum_{|I|=2^k} \sum_{J \subset I_1, |J|=2^{-n+k}} \|\Delta_J^\nu g\|_\nu^2 \right)^{1/2}.$$

As a result we get the following estimate for $r_{n,k}$:

$$r_{n,k} \leq C(K) \left(\sum_{|I|=2^k} \|\Delta_I^\mu f\|_\mu^2 \right)^{1/2} \left(\sum_{|J|=2^{-n+k}} \|\Delta_J^\nu g\|_\nu^2 \right)^{1/2}.$$

Now it is obvious from the formulae for T and $r_{n,k}$ that

$$T \le \sum_n 2^{-n/2} \sum_k r_{n,k} \, .$$

But from the estimate above and the Cauchy inequality, $\sum_k r_{n,k} \le C(K) \, \|f\|_\mu \|g\|_\nu$. So we proved Theorem 20.3.   □

CHAPTER 21

Difficult Terms and Several Paraproducts

Let us recall that f, g are good functions and that in the sum

$$(21.1) \qquad \tau := \sum_{|J| < 2^{-r}|I|, J \subset I, \operatorname{dist}(J, e(I)) \geq |I|^{3/4}|J|^{1/4}} (H_\mu \Delta_I^\mu f, \Delta_J^\nu g)_\nu .$$

we consider each term of τ and split it into three terms. To do this, let I_i denote the half of I that contains J. And I_n is another half. Let S denote the smallest superinterval of I in the same lattice: $S \in \mathcal{D}^\mu$, such that

$$(21.2) \qquad I \in \mathcal{O}_S ,$$

where the family of intervals $\mathcal{O}_S$ was introduced shortly after (20.12) as follows:

Notation. If $S \in \mathcal{D}^\mu$ is a stopping interval, and $\mathcal{S}_S = \{s\}, s \in \mathcal{D}^\mu$ is a collection of its maximal stopping subintervals, we call $\mathcal{O}_S$ the collection of all intervals I from *both* lattices $\mathcal{D}^\mu$, $\mathcal{D}^\nu$, such that the top side of the square Q_I lies in the set $\Omega_S := (\bar{Q}_S \setminus \cup_{s \in \mathcal{S}_S} \bar{Q}_s)$.

We wrote

$$(H_\mu \Delta_I^\mu f, \Delta_J^\nu g)_\nu = (H_\mu(\chi_{I_n} \Delta_I^\mu f), \Delta_J^\nu g)_\nu + (H_\mu(\chi_{I_i} \Delta_I^\mu f), \Delta_J^\nu g)_\nu$$
$$= (H_\mu(\chi_{I_n} \Delta_I^\mu f), \Delta_J^\nu g)_\nu + \langle \Delta_I^\mu f \rangle_{\mu, I_i} (H_\mu(\chi_S), \Delta_J^\nu g)_\nu$$
$$- \langle \Delta_I^\mu f \rangle_{\mu, I_i} (H_\mu(\chi_{S \setminus I_i}), \Delta_J^\nu g)_\nu .$$

Here $\langle \Delta_I^\mu f \rangle_{\mu, I_i}$ is the average of $\Delta_I^\mu f$ with respect to μ over I_i, which is the value of this function on I_i (by construction $\Delta_I^\mu f$ assumes on I two values, one on I_i, one on I_n).

The sum of absolute values of the first terms and the sum of absolute values of the third terms were already bounded by $C\|f\|_\mu \|g\|_\nu$ in the preceding sections. Middle terms were called "difficult terms", and we are going to estimate the absolute value of the sum of all difficult terms now (and not the sum of absolute values as before). This is the most difficult part of the proof.

Let $\{S\}_{S \in \mathcal{S}}$ denote the family of stopping intervals of all generations (for convenience we think that we are on the circle $\mathbb{T}$ and the first generation consists of the circle itself).

Notation. Let $S \in \mathcal{S}$ be an arbitrary stopping interval. We denote by $\mathbb{P}_{\mu, \mathcal{O}_S}$ the orthogonal projection in $L^2(\mu)$ onto the space generated by $\{h_I^\mu\}$, $I \in \mathcal{O}_S$, and we denote by $\mathbb{P}_{\nu, \mathcal{O}_S}$ the orthogonal projection in $L^2(\nu)$ onto the space generated by $\{h_J^\nu\}$, $J \in \mathcal{O}_S$. (Recall that $\mathcal{O}_S$ included by definition the intervals in both lattices $\mathcal{D}^\mu$ and $\mathcal{D}^\nu$.)

We fix $I \in \mathcal{D}^\mu$; it defines $S \in \mathcal{S}$ (see (21.2)). We look at terms

$$\langle \Delta_I^\mu f \rangle_{\mu, I_i} (H_\mu(\chi_S), \Delta_J^\nu g)_\nu .$$

We can write every term $\langle \Delta_I^\mu f \rangle_{\mu, I_i}(H_\mu(\chi_S), \Delta_J^\nu g)_\nu$ with fixed S and $I \in \mathcal{O}_S, J \in \mathcal{O}_S$ as

$$\langle \Delta_I^\mu \mathbb{P}_{\mu, \mathcal{O}_S} f \rangle_{\mu, I_i}(H_\mu(\chi_S), \Delta_J^\nu \mathbb{P}_{\nu, \mathcal{O}_S} g)_\nu \,.$$

The definition of τ_S . We collect all of these terms with $I \in \mathcal{O}_S, I \in \mathcal{D}^\mu, J \in \mathcal{O}_S$, $J \in \mathcal{D}^\nu, |J| \leq 2^{-r}|I|, J$ is good. The resulting sum is called τ_S. (In our summation below we should remember that f, g are good so we can sum over all pertinent pairs of I, J remembering that some of the Δ's are zero anyway.)

We first fix good J, then summing over such I's gives (such I's should contain J, and they form a "tower" of nested intervals, from the smallest one called $\ell(J)$ to the largest one equal to S; notice that the summing of quantities $\langle \Delta_I^\mu \varphi \rangle_{\mu, I}$ over such a "tower" results in the average over the smallest interval minus the average over the largest interval of the "tower", the latter one being zero in our case)

$$\langle \mathbb{P}_{\mu, \mathcal{O}_S} f \rangle_{\mu, \ell(J)}(\Delta_J^\nu H_\mu(\chi_S), \mathbb{P}_{\nu, \mathcal{O}_S} g)_\nu \,,$$

where $\ell(J) \in \mathcal{O}_S, \ell(J) \in \mathcal{D}^\mu, |\ell(J)| = 2^{r-1}|J|$. We denote $\mathcal{G}_S := \{I \in \mathcal{D}^\mu, I \in \mathcal{O}_S, I \text{ is good}\}$. Summing over J, we get

$$\tau_S = \sum_{I \in \mathcal{G}_S} \langle \mathbb{P}_{\mu, \mathcal{O}_S} f \rangle_{\mu, I} \Big(\sum_{J \in \mathcal{D}^\nu, J \in \mathcal{O}_S, |J|=2^{-r+1}|I|, J \text{ is good}} \Delta_J^\nu H_\mu(\chi_S), \mathbb{P}_{\nu, \mathcal{O}_S} g \Big)_\nu \,.$$

21.1. First paraproduct

Let us introduce our first paraproduct operator

$$\pi_{H_\mu \chi_S} \varphi := \sum_{I \in \mathcal{D}^\mu, I \in \mathcal{O}_S, I \text{ is good}} \langle \varphi \rangle_{\mu, I} \sum_{J \in \mathcal{D}^\nu, J \in \mathcal{O}_S, J \subset I, |J|=2^{-r+1}|I|, J \text{ is good}} \Delta_J^\nu H_\mu(\chi_S) \,.$$

Then the absolute value of the sum τ_S above is

$$(21.3) \qquad |(\pi_{H_\mu \chi_S} \mathbb{P}_{\mu, \mathcal{O}_S} f, \mathbb{P}_{\nu, \mathcal{O}_S} g)_\nu| \leq C_1 \|\mathbb{P}_{\mu, \mathcal{O}_S} f\|_\mu \|\mathbb{P}_{\nu, \mathcal{O}_S} g\|_\nu \,,$$

where C_1 is the norm of $\pi_{H_\mu \chi_S}$ as an operator from $L^2(\mu)$ to $L^2(\nu)$.

THEOREM 21.1. *The norm of operator $\pi_{H_\mu \chi_S}$ as an operator from $L^2(\mu)$ to $L^2(\nu)$ is bounded by $C_1(K) < \infty$, where K is the constant participating in the definition of stopping intervals.*

Proof. Obviously, the orthogonality in $L^2(\nu)$ of $\Delta_J^\nu H_\mu(\chi_S)$ gives

$$\|\pi_{H_\mu \chi_S} \varphi\|_\nu^2 \leq \sum_{I \in \mathcal{D}^\mu, I \in \mathcal{O}_S} |\langle \varphi \rangle_{\mu, I}|^2 a_I,$$

where $\Phi(I) := \{J : J \in \mathcal{D}^\nu, J \in \mathcal{O}_S, J \subset I, |J| = 2^{-r+1}|I|, \text{dist}(J, \partial(I)) \geq |I|^{3/4}|I|^{1/4}\}$

$$a_I := \sum_{J \in \Phi(I)} \|\Delta_J^\nu H_\mu(\chi_S)\|_\nu^2 \,.$$

The Carleson imbedding theorem (see [21], and in this context [43]) says that the boundedness of the sum $\sum_{I \in \mathcal{D}^\mu, I \in \mathcal{O}_S} |\langle \varphi \rangle_{\mu, I}|^2 a_I$ by $C \|\varphi\|_\mu^2$ is equivalent to the following Carleson condition:

$$(21.4) \qquad \forall I \in \mathcal{D}^\mu, I \in \mathcal{O}_S \sum_{\ell \in \mathcal{D}^\mu, \ell \in \mathcal{O}_S, \ell \subset I} a_\ell \leq c \, \mu(I) \,.$$

Of course ($\Psi(I) := \{J : J \in \mathcal{D}^\nu, J \in \mathcal{O}_S, J \subset I, |J| \le 2^{-r+1}|I|, \operatorname{dist}(J, \partial(I)) \ge |I|^{3/4}|I|^{1/4}\}$)

$$\sum_{\ell \in \mathcal{D}^\mu, \ell \in \mathcal{O}_S, \ell \subset I} a_\ell = \sum_{J : J \in \Psi(I)} \|\Delta_J^\nu H_\mu(\chi_S)\|_\nu^2 = \Big\| \sum_{J : J \in \Psi(I)} \Delta_J^\nu H_\mu(\chi_S) \Big\|_\nu^2.$$

By duality then

$$\sum_{\ell \in \mathcal{D}^\mu, \ell \in \mathcal{O}_S, \ell \subset I} a_\ell = \sup_{\psi \in L^2(\nu), \|\psi\|_\nu = 1} \Big| \sum_{J : J \in \Psi(I)} (H_\mu(\chi_S), \Delta_J^\nu \psi)_\nu \Big|^2$$

$$\le \sup_{\psi \in L^2(\nu), \|\psi\|_\nu = 1} \Big| \sum_{J : J \in \Psi(I)} (H_\mu(\chi_{S \setminus I}), \Delta_J^\nu \psi)_\nu \Big|^2 + \|H_\mu(\chi_I)\|_\nu^2.$$

So (15.14) implies

$$(21.5) \quad \sum_{\ell \in \mathcal{D}^\mu, \ell \in \mathcal{O}_S, \ell \subset I} a_\ell \le \sup_{\psi \in L^2(\nu), \|\psi\|_\nu = 1} \Big| \sum_{J : J \in \Psi(I)} (H_\mu(\chi_{S \setminus I}), \Delta_J^\nu \psi)_\nu \Big|^2 + C_\chi \, \mu(I).$$

Let us consider the term $(H_\mu(\chi_{S \setminus I}), \Delta_J^\nu \psi)_\nu$, $J \in \Psi(I)$. Exactly this quantity was estimated in (20.5). We get

$$|(H_\mu(\chi_{S \setminus I}), \Delta_J^\nu \psi)_\nu| \le A\,\nu(J)^{1/2} \|\Delta_J^\nu \psi\|_\nu \Big(\frac{|J|}{|I|}\Big)^{1/2} P_I(\chi_{S \setminus I}) \, d\mu.$$

So the first term in (21.5) is bounded by (we use the Cauchy inequality)

$$\sum_{J : J \in \Psi(I)} \frac{|J|}{|I|} [P_I(\chi_{S \setminus I}) \, d\mu]^2 \nu(J) \le \sum_n 2^{-n} \sum_{|J| = 2^{-n}|I|, J \subset I} [P_I(\chi_{S \setminus I}) \, d\mu]^2 \nu(J)$$

$$= \sum_n 2^{-n} [P_I(\chi_{S \setminus I}) \, d\mu]^2 \nu(I)$$

as $\|\psi\|_\nu = 1$. It is time to use the fact that $I \in \mathcal{O}_S$, which means that the stopping criterion (20.7) is not yet achieved on I; in other words,

$$[P_I(\chi_{S \setminus I}) \, d\mu]^2 \nu(I) \le K \, \mu(I).$$

Combining this with (21.5) we get (21.4):

$$\sum_{\ell \in \mathcal{D}^\mu, \ell \in \mathcal{O}_S, \ell \subset I} a_\ell \le (K + C_\chi) \, \mu(I),$$

and Theorem 21.1 is proved. $\qquad\square$

Let us recall that we introduced above the definition of τ_S, for stopping interval S. We finished the estimate of the sum of τ_S over all stopping S (recall that the set of all stopping intervals was called $\mathcal{S}$):

$$(21.6) \quad \sum_{S \in \mathcal{S}} \tau_S \le C(K, C_\chi) \sum_{S \in \mathcal{S}} \|\mathbb{P}_{\mu, \mathcal{O}_S} f\|_\mu \|\mathbb{P}_{\nu, \mathcal{O}_S} g\|_\nu \le C(K, C_\chi) \|f\|_\mu \|g\|_\nu,$$

the last inequality following from the orthogonality of $\mathbb{P}_{\mu, \mathcal{O}_S} f$ for different S (the same for $\mathbb{P}_{\nu, \mathcal{O}_S} g$) and the Cauchy inequality. The orthogonality follows from the definition of $\mathcal{O}_S$, which implies the disjointedness of $\mathcal{O}_S$ for different S.

21.2. Two more paraproducts

In the previous section we have estimated a piece of the sum of the difficult terms

$$(21.7) \qquad \langle \Delta_I^\mu f \rangle_{\mu, I_i} (H_\mu(\chi_S), \Delta_J^\nu g)_\nu \, ;$$

namely, we estimated the sum of such terms, when I, J both lie in the same family $\mathcal{O}_S$, where $S \in \mathcal{S}$ (arbitrary stopping interval). Such a sum was called τ_S, and we just proved in (21.6) that $\sum_{S \in \mathcal{S}} \tau_S \le C \|f\|_\mu \|g\|_\nu$.

What is left is to estimate the sum of abovementioned terms when $J \in \mathcal{O}_S$ and I belongs to another $\mathcal{O}_{\hat{S}}$, where $S, \hat{S}$ are both stopping intervals. As I is larger than J, we have to consider the pairs of stopping intervals where S is strictly inside $\hat{S}$ ($\hat{S}$ is one or more generations higher than S).

Let us fix J. Let $\cdots \subset S_2 \subset S_1 \subset S_0 \subset \ldots$ be a (finite) sequence of stopping intervals of successive generations containing J. The sequence for I's, over which we have to sum up, will be one term shorter (the smallest one should be discarded). This is because we sum up all the terms, where J and I are in different families $\mathcal{O}_S, \mathcal{O}_{\hat{S}}$, and S is smaller. Notice also that $\langle \Delta_I^\mu f \rangle_{\mu, I_i}$ is the difference betweentwo averages of f with respect to μ, one over I_i and one over its father I. It is easy to sum up successive differences and summing all above mentioned terms with fixed J, we get

$$\cdots + (\langle f \rangle_{\mu, S_1} - \langle f \rangle_{\mu, S_0})(H_\mu \chi_{S_0}, \Delta_J^\nu g)_\nu + (\langle f \rangle_{\mu, S_2} - \langle f \rangle_{\mu, S_1})(H_\mu \chi_{S_1}, \Delta_J^\nu g)_\nu + \cdots$$

Regrouping, we get

$$\cdots + \langle f \rangle_{\mu, S_1}((H_\mu \chi_{S_0 \setminus S_1}, \Delta_J^\nu g)_\nu + \cdots .$$

We can ignore the terms with the largest S_M for a given J, for which there will be no pair. This is because the largest S_M is always the unit circle $\mathbb{T}$. We have made a convention that we are on the circle rather than on the real line. On the other hand, the average of f over the circle was assumed to vanish. We assumed this at the beginning of the proof of our main theorem, and explained there why this does not lose generality.

But we have to take into consideration the terms with the smallest S_m for a given J, for which there will be no pair. Subsequently, the sum of the abovementioned terms in (21.7), when $J \in \mathcal{O}_S$ and I belongs to another $\mathcal{O}_{\hat{S}}$ (where $S, \hat{S}$ are both stopping intervals, and S is strictly smaller than $\hat{S}$) can be written in the following form.

We denote below by $\hat{S}$ the stopping interval containing the stopping S and of the previous generation (the stopping father of S), and by D_S the $\cup_{s \in \mathcal{S}, s \subset S, s \neq S} \mathcal{O}_s$, that is the union of $\mathcal{O}_s$ over the stopping children, grandchildren, et cetera of S.

$$(21.8) \qquad \rho := \sum_{s \in \mathcal{S}} \langle f \rangle_{\mu, S}(H_\mu \chi_{\hat{S} \setminus S}, \sum_{J \in D_S} \Delta_J^\nu g)_\nu + \sum_{s \in \mathcal{S}} \langle f \rangle_{\mu, S}(H_\mu \chi_{\hat{S}}, \sum_{J \in \mathcal{O}_S} \Delta_J^\nu g)_\nu \, .$$

We can rewrite

$$\rho = \sum_{s \in \mathcal{S}} \langle f \rangle_{\mu, S}(H_\mu \chi_{\hat{S} \setminus S}, \sum_{J \in D_S \cup \mathcal{O}_S} \Delta_J^\nu g)_\nu + \sum_{s \in \mathcal{S}} \langle f \rangle_{\mu, S}(H_\mu \chi_S, \sum_{J \in \mathcal{O}_S} \Delta_J^\nu g)_\nu \, .$$

We use now the notation $Q_S := D_S \cup \mathcal{O}_S$; that is, $Q_S = \cup_{s \in \mathcal{S}, s \subset S} \mathcal{O}_s$. This means that the family Q_S consists of intervals ℓ from both our lattices $\mathcal{D}^\mu, \mathcal{D}^\nu$ such that the top of the square Q_ℓ belongs to the square Q_S (the slight abuse of notation, the square, and the corresponding family of the intervals are denoted by the same letter). Then we can introduce two projections $\mathbb{P}_{\nu,Q_S}$, $\mathbb{P}_{\nu,\mathcal{O}_S}$. Actually the second one was already introduced. But anyway, we denote by $\mathbb{P}_{\nu,\mathcal{O}_S}$ the orthogonal projection in $L^2(\nu)$ onto the space generated by $\{h_J^\nu\}$, $J \in \mathcal{O}_S$ with good J. And we denote by $\mathbb{P}_{\nu,Q_S}$ the orthogonal projection in $L^2(\nu)$ onto the space generated by $\{h_J^\nu\}$, $J \in Q_S$ with good J. (Recall that $Q_S, \mathcal{O}_S$ included by definition the intervals from both lattices $\mathcal{D}^\mu$ and $\mathcal{D}^\nu$.) Now we can write ρ as follows:

$$\rho = \sum_{s \in \mathcal{S}} \langle f \rangle_{\mu,S} (H_\mu \chi_{\hat{S} \setminus S}, \mathbb{P}_{\nu,Q_S} g)_\nu + \sum_{s \in \mathcal{S}} \langle f \rangle_{\mu,S} (H_\mu \chi_S, \mathbb{P}_{\nu,\mathcal{O}_S} g)_\nu =: \rho_1 + \rho_2 \,.$$

We introduce now two paraproducts

$$\pi^{\mathcal{O}} := \sum_{s \in \mathcal{S}} \langle f \rangle_{\mu,S} \mathbb{P}_{\nu,\mathcal{O}_S} (H_\mu \chi_S) \,,$$

$$\pi^{Q} := \sum_{s \in \mathcal{S}} \langle f \rangle_{\mu,S} \mathbb{P}_{\nu,Q_S} (H_\mu \chi_{\hat{S} \setminus S}) \,.$$

Then $\rho_1 = (\pi^{\mathcal{O}} f, g)_\nu, \rho_2 = (\pi^{Q} f, g)_\nu$. So to finish the proof of our main Theorem 15.1, it is enough to prove the boundedness of these paraproducts as operators from $L^2(\mu)$ to $L^2(\nu)$.

To prove the boundedness of the first paraproduct let us use Theorem 20.1. Consider the sequence

$$\{a_S\}_{S \in \mathcal{S}}, \; a_S := \|\mathbb{P}_{\nu,\mathcal{O}_S} (H_\mu \chi_S)\|_\nu^2 \,.$$

It is a Carleson sequence:

$$(21.9) \qquad \forall I \in \mathcal{D}^\mu \quad \sum_{S \subset I, S \in \mathcal{S}} a_S \leq C \, \mu(I) \,.$$

In fact, $a_S \leq \|H_\mu \chi_S\|_\nu^2 \leq C_\chi \, \mu(S)$ by (15.14). Now (21.9) becomes clear by Theorem 20.1.

Notice that $\mathbb{P}_{\nu,\mathcal{O}_S}$ are mutually orthogonal projections in $L^2(\nu)$ for different S. This is just because the families $\mathcal{O}_S$ are pairwise disjoint for different $S \in \mathcal{S}$. We already saw this type of paraproduct with the property of orthogonality (see [21], [43], and especially Theorem 21.1 above). And we know that Carleson condition (21.9) is sufficient for the paraproduct operator $\pi^{\mathcal{O}}$ to be bounded.

The second paraproduct π^{Q}. The main problem is that $\mathbb{P}_{\nu,Q_S}$ are *not mutually orthogonal* projections in $L^2(\nu)$.

So $\|\pi^{Q} f\|_\nu^2$ has the diagonal part but also the off-diagonal part:

$$DP := \sum_{S \in \mathcal{S}} |\langle f \rangle_{\mu,S}|^2 \|\mathbb{P}_{\nu,Q_S} H_\mu(\chi_{\hat{S} \setminus S})\|_\nu^2 \,,$$

$$ODP := \sum_{S,S' \in \mathcal{S}, S' \subset S, S' \neq S} |\langle f \rangle_{\mu,S'}| |\langle f \rangle_{\mu,S}| |(\mathbb{P}_{\nu,Q_S} H_\mu(\chi_{\hat{S} \setminus S}), \mathbb{P}_{\nu,Q_{S'}} H_\mu(\chi_{\hat{S}' \setminus S'}))_\nu|$$

$$= \sum_{S,S' \in \mathcal{S}, S' \subset S, S' \neq S} |\langle f \rangle_{\mu,S'}| |\langle f \rangle_{\mu,S}| |(\mathbb{P}_{\nu,Q_{S'}} H_\mu(\chi_{\hat{S} \setminus S}), \mathbb{P}_{\nu,Q_{S'}} H_\mu(\chi_{\hat{S}' \setminus S'}))_\nu| \,.$$

We start with the *ODP*. Let $r = r(S', S)$ be the generation gap between S' and S. Then

$$ODP \leq \sum_{S,S' \in \mathcal{S}, S' \subset S, S' \neq S} |\langle f \rangle_{\mu,S}|^2 \|\mathbb{P}_{\nu,Q_{S'}} H_\mu(\chi_{\hat{S} \setminus S})\|_\nu^2 \cdot (1+\varepsilon)^{r(S',S)}$$

$$+ \sum_{S,S' \in \mathcal{S}, S' \subset S, S' \neq S} |\langle f \rangle_{\mu,S'}|^2 \|\mathbb{P}_{\nu,Q_{S'}} H_\mu(\chi_{\hat{S}' \setminus S'})\|_\nu^2 \cdot (1+\varepsilon)^{-r(S',S)}$$

$$\leq \sum_{S \in \mathcal{S}} |\langle f \rangle_{\mu,S}|^2 \sum_{j=1}^{\infty} (1+\varepsilon)^j \sum_{S' \in \mathcal{S}, S' \subset S, r(S',S)=j} \|\mathbb{P}_{\nu,Q_{S'}} H_\mu(\chi_{\hat{S} \setminus S})\|_\nu^2$$

$$+ C(\varepsilon) \sum_{S \in \mathcal{S}} |\langle f \rangle_{\mu,S}|^2 \|\mathbb{P}_{\nu,Q_S} H_\mu(\chi_{\hat{S} \setminus S})\|_\nu^2 .$$

Now we need to estimate these sums:

$$F_j := \sum_{S \in \mathcal{S}} |\langle f \rangle_{\mu,S}|^2 \sum_{S' \in \mathcal{S}, S' \subset S, r(S',S)=j} \|\mathbb{P}_{\nu,Q_{S'}} H_\mu(\chi_{\hat{S} \setminus S})\|_\nu^2 , \quad j = 1, 2, 3, \ldots ,$$

$$F_0 := \sum_{S \in \mathcal{S}} |\langle f \rangle_{\mu,S}|^2 \|\mathbb{P}_{\nu,Q_S} H_\mu(\chi_{\hat{S} \setminus S})\|_\nu^2 .$$

By the way, $F_0 = DP$.

All such sums have the form of Carleson imbedding theorems. So we need to check the countable number of Carleson conditions now.

Carleson condition for F_j. We introduce the sequence

$$a_S := \|\mathbb{P}_{\nu,Q_S} H_\mu(\chi_{\hat{S} \setminus S})\|_\nu^2 , \quad S \in \mathcal{S} .$$

And also

$$a_S^j := \|\mathbb{P}_{\nu,Q_S} H_\mu(\chi_{\hat{S} \setminus \tilde{S}})\|_\nu^2 , \quad S, \hat{S}, \tilde{S} \in \mathcal{S}, \; r(S, \tilde{S}) = j, r(\tilde{S}, \hat{S}) = 1, \; j = 1, 2, 3, \ldots .$$

We first establish a Carleson property for $\{a_S\}$. By Theorem 20.1 it is enough to fix a stopping $\hat{S}$ and to prove

$$(21.10) \qquad \sum_{S \in \mathcal{S}, S \subset \hat{S}, r(S,\hat{S})=1} \|\mathbb{P}_{\nu,Q_S} H_\mu(\chi_{\hat{S} \setminus S})\|_\nu^2 \leq C \mu(\hat{S}) .$$

Let $\|\psi\|_\nu = 1$. Let us consider the term $(H_\mu(\chi_{\hat{S} \setminus S}), \Delta_J^\nu \psi)_\nu$, $J \in Q_S$. Exactly this quantity was estimated in (20.5). We get

$$|(H_\mu(\chi_{\hat{S} \setminus S}), \Delta_J^\nu \psi)_\nu| \leq A \nu(J)^{1/2} \|\Delta_J^\nu \psi\|_\nu \left(\frac{|J|}{|S|} \right)^{1/2} P_S(\chi_{\hat{S} \setminus S}) \, d\mu .$$

So each term can be estimated as follows:

$$(21.11) \qquad \|\mathbb{P}_{\nu,Q_S} H_\mu(\chi_{\hat{S} \setminus S})\|_\nu^2 \leq (P_S(\chi_{\hat{S} \setminus S}) \, d\mu)^2 \sum_{J \text{ good}, J \subset S} \nu(J) \frac{|J|}{|S|} .$$

So the sum in (21.10) is bounded by

$$\sum_{S \in \mathcal{S}, S \subset \hat{S}, r(S,\hat{S})=1} (P_S(\chi_{\hat{S} \setminus S}) \, d\mu)^2 \sum_{J \text{ good}, J \subset S} \nu(J) \frac{|J|}{|S|}$$

$$\leq \sum_{S \in \mathcal{S}, S \subset \hat{S}, r(S,\hat{S})=1} (P_S(\chi_{\hat{S} \setminus S}) \, d\mu)^2 \nu(S) \leq K \sum_{S \in \mathcal{S}, S \subset \hat{S}, r(S,\hat{S})=1} \mu(S) \leq K \mu(\hat{S}) ,$$

which proves (21.10). This gives

(21.12) $$DP = F_0 \le C \left\| f \right\|_\mu^2 \,.$$

21.3. Second paraproduct: miraculous improvement of the Carleson property

Now we start to investigate the Carleson property of $\{a_S^j\}_{S \in \mathcal{S}}$. Again Theorem 20.1 shows that it is sufficient to prove that for a given stopping interval $\hat{S}$,

(21.13) $$\sum_{\tilde{S} \in \mathcal{S}, \tilde{S} \subset \hat{S}, r(\tilde{S}, \hat{S}) = 1} \sum_{S \in \mathcal{S}, S \subset \tilde{S}, r(S, \tilde{S}) = j} \left\| \mathbb{P}_{\nu, Q_S}(H_\mu(\chi_{\hat{S} \setminus \tilde{S}})) \right\|_\nu^2 \le C \, 2^{-j} \mu(\hat{S}) \,.$$

If this holds, then we can choose ε sufficiently small in the estimate of ODP above in such a way that

(21.14) $$\sum F_j \le C \left\| f \right\|_\mu^2 \,.$$

And the estimate of ODP would be done. Combining (21.12) and (21.14), we finish our two–weight estimate completely.

But first we need to prove (21.13). So consider $S, \tilde{S}, \hat{S}$ as in (21.13). Let $\left\| \psi \right\|_\nu = 1$. Let us consider the term $(H_\mu(\chi_{\hat{S} \setminus \tilde{S}}), \Delta_J^\nu \psi)_\nu$, $J \in Q_S$. Exactly this quantity was estimated in (20.5). We get

$$\left| (H_\mu(\chi_{\hat{S} \setminus \tilde{S}}), \Delta_J^\nu \psi)_\nu \right| \le A \, \nu(J)^{1/2} \left\| \Delta_J^\nu \psi \right\|_\nu \left(\frac{|J|}{|\tilde{S}|} \right)^{1/2} P_{\tilde{S}}(\chi_{\hat{S} \setminus \tilde{S}}) \, d\mu \,.$$

As in (21.11), we can write

(21.15) $$\left\| \mathbb{P}_{\nu, Q_S} H_\mu(\chi_{\hat{S} \setminus \tilde{S}}) \right\|_\nu^2 \le (P_{\tilde{S}}(\chi_{\hat{S} \setminus \tilde{S}}) \, d\mu)^2 \sum_{J \text{ good}, J \subset S} \nu(J) \frac{|J|}{|\tilde{S}|} \,.$$

But between J and $\tilde{S}$ there are at least j dyadic generations! So we can continue (21.15) as follows:

$$\left\| \mathbb{P}_{\nu, Q_S} H_\mu(\chi_{\hat{S} \setminus \tilde{S}}) \right\|_\nu^2 \le (P_{\tilde{S}}(\chi_{\hat{S} \setminus \tilde{S}}) \, d\mu)^2 \nu(S) \sum_{t = j}^{\infty} \sum_{J \subset S, |J| = 2^{-t} |\tilde{S}|} \frac{|J|}{|\tilde{S}|} \,.$$

Therefore,

(21.16) $$\left\| \mathbb{P}_{\nu, Q_S} H_\mu(\chi_{\hat{S} \setminus \tilde{S}}) \right\|_\nu^2 \le A \, 2^{-j} (P_{\tilde{S}}(\chi_{\hat{S} \setminus \tilde{S}}) \, d\mu)^2 \nu(S) \,.$$

Now the improved Carleson condition (21.13) follows from this last inequality as a consequence of the stopping condition (20.5).

Theorem 15.1 is completely proved.

CHAPTER 22

Two-Weight Hilbert Transform and Maximal Operator

22.1. Doubling

Let us introduce two maximal operators.

$$M_\mu f(x) := \sup_{I:x\in I} \frac{1}{|I|} \int_I |f| \, d\mu, \ M_\nu g(x) := \sup_{I:x\in I} \frac{1}{|I|} \int_I |g| \, d\nu \, .$$

By the works of E. Sawyer [58], [59], it is known when M_μ is a bounded operator from $L^2(\mu)$ to $L^2(\nu)$. This happens if and only if the uniform bound on test functions holds:

$$(22.1) \qquad \|M_\mu \chi_I\|_\nu^2 \leq C_m \, \mu(I), \ \forall \text{ interval } I \, .$$

The symmetric condition (with exchanging μ and ν) is necessary and sufficient for the boundedness of M_ν:

$$(22.2) \qquad \|M_\nu \chi_I\|_\mu^2 \leq C_m \, \nu(I), \ \forall \text{ interval } I \, .$$

Of course, (15.14), (15.15) of our Theorem (15.1):

$$(22.3) \qquad \|H_\mu \chi_I\|_{L^2(\nu)}^2 \leq C_\chi \nu(I), \ \forall I \subset \mathbb{R} \, ,$$

$$(22.4) \qquad \|H_\nu \chi_I\|_{L^2(\mu)}^2 \leq C_\chi \mu(I), \ \forall I \subset \mathbb{R} \, .$$

are the analogs of these Sawyer conditions, but applied to a singular operator.

The drawback of our main result about the two–weight Hilbert transform is that we were obliged to assume the "smoothness" of measures to get the necessary and sufficient conditions for the boundedness of $H_\mu : L^2(\mu) \to L^2(\nu)$. Smoothness meant that the support of the measure is the whole line (or circle) and the doubling property holds.

The next theorem does not have the smoothness requirement but it has still the doubling requirement. It is now understood as

$$(22.5) \qquad \mu(2I) \leq C_d \mu(I), \text{ if } \mu(I) > 0, \ \nu(2I) \leq C_d \nu(I), \text{ if } \nu(I) > 0 \, .$$

The theorem below gives the necessary and sufficient conditions not for the boundedness of $H_\mu : L^2(\mu) \to L^2(\nu)$, but for the boundedness of the *family* consisting of three operators: H_μ, M_μ, M_ν. Such *family* results can be found in the literature. See, for example, [50], where the two–weight criterion has been found for the family of all Martingale Transforms.

THEOREM 22.1. *If operators H_μ, M_μ are bounded from $L^2(\mu)$ to $L^2(\nu)$, and M_ν is bounded from $L^2(\nu)$ to $L^2(\mu)$, then the constants C_m, C_χ in (22.1), (22.2), (22.3), (22.4), and the constant C_p from (15.16) are finite and bounded by $A \, (\|H_\mu\|_{L^2(\mu)\to L^2(\nu)} + \|M_\mu\|_{L^2(\mu)\to L^2(\nu)} + \|M_\nu\|_{L^2(\nu)\to L^2(\mu)})^2$.*

On the other hand, let (22.5) be satisfied; then the norms of these three operators are bounded by $C(C_d, C_m, C_\chi, C_p) < \infty$, if the constants C_d, C_m, C_χ, C_p are finite. In other words, the family H_μ, M_μ, M_ν consists of bounded operators if and only if test conditions (22.1), (22.2), (22.3), (22.4), and (15.16) are satisfied.

Remark. It only seems that the theorem is asymmetric. Of course, the boundedness of H_μ from $L^2(\mu)$ to $L^2(\nu)$ is the same as the boundedness of minus its adjoint H_ν from $L^2(\nu)$ to $L^2(\mu)$. Note also that the essence of the theorem is that the norm $\|H_\mu\|_{L^2(\mu) \to L^2(\nu)}$ can be bounded by $C(C_d, C_m, C_\chi, C_p)$. The rest is either obvious or follows from Sawyer's two–weight result for maximal functions cited above.

Proof. We repeat verbatim all lines of the proof of Theorem 15.1. Again introduce good and bad functions, and reduce by averaging over probability the estimate of H_μ to its estimate on good functions. Everything goes without change up to the moment when we have to prove Theorem 20.1. There the smoothness was used; we do not have it now.

So here are the changes in the proof of Theorem 15.1. First of all, we change a bit the stopping criterion (20.7). Now, given $\hat{I} \in \mathcal{D}^\mu$, we (fixing a large K) consider the largest $I \subset \hat{I}, I \in \mathcal{D}^\mu$ such that

$$(22.6) \qquad \left[P_I(\chi_{\hat{I}} \, d\mu) \right]^2 \nu(I) \geq K \, \mu(I), \ \mu(I) > 0 \,.$$

Call the family of such I's by $\mathcal{F}(\hat{I})$. Then we call the interval $S \subset \hat{I}, S \in \mathcal{D}^\mu$ a *stopping interval* if either it is a grandfather of such an $I \in \mathcal{F}(\hat{I})$ or $\mu(S) = 0$.

Here is the place, where we use the finiteness of C_m instead of the smoothness of measure property:

LEMMA 22.2. *If C_m, C_d are finite, then for every $\hat{I} \in \mathcal{D}^\mu$,*

$$(22.7) \qquad \sum_{S \in \mathcal{D}^\mu, \, S \subset \hat{I}, \, S \text{ is maximal stopping}} \mu(S) \leq \frac{1}{2} \mu(\hat{I}) \,,$$

provided that the constant K in the stopping criterion (22.6) is large enough.

Proof. It is a standard estimate of the Poisson integral via the maximal function (see, for example, [**21**]), which gives

$$(22.8) \qquad P_I(\chi_{\hat{I}} \, d\mu) \leq A \inf_{x \in I} (M_\mu \chi_{\hat{I}})(x) \,.$$

Then (22.8) implies

$$K \sum_{I: \, I \in \mathcal{F}(\hat{I}), I \subset \hat{I}} \mu(I) \leq \sum_{I: \, I \in \mathcal{F}(\hat{I}), I \subset \hat{I}} P_I(\chi_{\hat{I}} \, d\mu)^2 \nu(I)$$

$$\leq \int_{\cup I \, I \in \mathcal{F}(\hat{I}), I \subset \hat{I}} (M_\mu \chi_{\hat{I}})(x)^2 \, d\nu(x) \leq \int (M_\mu \chi_{\hat{I}})(x)^2 \, d\nu(x) \leq C_m \mu(\hat{I}) \,.$$

If S is a maximal stopping interval inside $\hat{I}$, then either $\mu(S) = 0$ or it contains a grandson $I \in \mathcal{F}(\hat{I})$ of positive measure. Use (22.5). Then (22.7) is proved if K is larger than $A \, C_d \, C_m$. $\qquad \square$

We construct stopping intervals of the next generation inside stopping intervals already constructed, and we continue this process. We obtained the family of all stopping intervals of all generations. Call it $\mathcal{S}$ as before.

Let $\hat{I}$ be a stopping interval, $\mu(\hat{I}) > 0$, and let $I \subset \hat{I}, I \in \mathcal{D}^\mu$ be such that $\hat{I}$ is the smallest interval from $\mathcal{S}$ containing I; then, without loss of generality, we can consider only the case $\mu(I) > 0$ (otherwise any $f \in L^2(\mu)$ does not live on I, and we do not need to consider $(\Delta^\mu_I f, \dots)_\nu$ for $I \subset \hat{I}$). Let now $I_i, i = 1, 2$ be two halves of I; then, by construction,

$$(22.9) \qquad \left[P_{I_i}(\chi_{\hat{I}\setminus I_i}\, d\mu) \right]^2 \nu(I) \le K\, \mu(I_i), \quad i = 1, 2\,.$$

The relation (22.9) is the only one that was used in the estimate of stopping terms.

The estimate of paraproducts is verbatim the same as in Section 21.2. Again we have the right to consider only those $S \in \mathcal{S}$ for which $\mu(S) > 0$. For such S, one has

$$(22.10) \qquad \left[P_S(\chi_{\hat{S}\setminus S}\, d\mu) \right]^2 \nu(S) \le K\, \mu(S)\,,$$

and this is the only thing we used for the estimates of paraproducts. $\qquad\square$

22.2. No doubling

THEOREM 22.3. *If operators H_μ, M_μ are bounded from $L^2(\mu)$ to $L^2(\nu)$, and M_ν is bounded from $L^2(\nu)$ to $L^2(\mu)$, then the constants C_m, C_χ in (22.1), (22.2), (22.3), (22.4), and the constant C_p from (15.16) are finite and bounded by $A\left(\|H_\mu\|_{L^2(\mu)\to L^2(\nu)} + \|M_\mu\|_{L^2(\mu)\to L^2(\nu)} + \|M_\nu\|_{L^2(\nu)\to L^2(\mu)}\right)^2$.*

On the other hand, the norms of these three operators are bounded by constants depending on C_m, C_χ, C_p, if the constants C_m, C_χ, C_p are finite. In other words, the family H_μ, M_μ, M_ν consists of bounded operators if and only if test conditions (22.1), (22.2), (22.3), and (22.4) are satisfied.

The difference with Theorem 22.1 is in the absence of the doubling assumption on the measures. The proof is slightly more complicated because the stopping time should be chosen more carefully. We will provide the proof elsewhere.

Bibliography

[1] J. Bourgain, *Some remarks on Banach spaces in which martingale difference sequences are unconditional*, Ark. Mat. , **21**, (1983), pp. 163–168.

[2] _____, *Vector-valued singular integrals and the $H^1 - BMO$ duality*, Probability Theory and Harmonic Analysis (Cleveland, Ohio 1983), Monographs and textbooks in Pure and Applied Mathematics, Dekker, New York, 1986.

[3] D. L. Burkholder, *A geometrical condition that implies the existence of certain singular integrals of Banach-space-valued functions,* Proc. Conf. Harmonic Analysis in honor of Antoni Zygmund, ed. W. Beckner, A. P. Calderón, R. Fefferman, and P. W. Jones, Wadsworth, Belmont, Ca., 1983.

[4] N. E. Benamara and N. K. Nikolski, *Resolvent tests for similarity to a normal operator*, Proc. London Math. Soc., **78**, (1999), no. 3, pp. 585–626.

[5] M. Christ, *A T(b) theorem with remarks on analytic capacity and the Cauchy integral*, Colloquium Math., v. LX/LXI, (1990), 601–628.

[6] M. Cotlar and C. Sadosky, *A moment theory approach to the Riesz theorem on the conjugate functions with general measures*, Studia Math., **53** (1975), no. 1, 75-101.

[7] _____, *Characterization of two measures satisfying the Riesz inequality for the Hilbert transform in L^2*, Acta Cient. Venezolana **30** (1979), no. 4, 346–348.

[8] _____, *On the Helson-Szegö theorem and a related class of modified Toeplitz kernels*, Harmonic Analysis in Euclidean spaces, Proc. Symp. Pure Math., v. 35, Part 1, Amer. Math. Soc., Providence, RI, 1979, pp. 383–407.

[9] _____, *Majorized Toeplitz forms and weighted inequalities with general norms*, Lecture Notes Math., vol. 908, Springer-Verlag, Berlin–Heidelberg–New York, 1982, pp. 139–168.

[10] _____, *On some L^p version of Helson-Szegó theorem*, Conference on harmonic analysis in honor of Antoni Zygmund, Wadsworth Math. Series, 1983, pp. 306–317.

[11] R. Coifman, P. Jones, and S. Semmes, *Two elementary proofs of the L^2 boundedness of the Cauchy integrals on Lipschitz curves*, J. Amer. Math. Soc. **2** (1989), no. 3, pp. 553–564.

[12] G. David and J.-L. Journé, *A boundedness criterion for generalized Calderón–Zygmund operators*, Annals of Math., **120**, (1984), no. 2, pp. 371–397.

[13] _____, *Une caracterisation des opérateurs intégraux singuliers bornés sur $L^2(\mathbb{R}^n)$*, Comptes Rendus Acad. Sci. Paris, **296**, (1983), no. 18, pp. 761–764.

[14] G. David, J.-L. Journé, and S. Semmes, *Opérateurs de Calderón–Zygmund, fonctions para-accretives et interpolation.* Rev. Mat. Iberoameriricana, **1**, (1985), no. 4, pp. 1–56.

[15] G. David, *Une nouvelle démonstration du théorème T(b), d'après Coifman et Semmes*, Lecture Notes Math., vol. 1438, Springer-Verlag, Berlin–Heidelberg–New York, 1990, pp. 39–50.

[16] _____, *Opérateurs intégraux singuliers sur certaines courbes du plan complexe*, Ann. Sci. Ecole Norm. Sup., **17**, (1984), 157–189.

[17] _____, *Analytic capacity, Calderón-Zygmund operators, and rectifiability*, Publ. Mat., v. 43, (1999), 3–25.

[18] G. David and P. Mattila, *Removable sets for Lipschitz harmonic functions in the plane*, Revista Mat. Iberoamericana, v. 16, (2000), 137–215.

[19] G. David, *Unrectifiable 1-sets have vanishing analytic capacity*, Revista Mat. Iberoamericana, v. 14, (1998), 369–479.

[20] H. Farag, *The Riesz kernels do not give rise to higher dimensional analogues of Menger-Melnikov curvature*, Publ. Mathemàtiques, v. 43, (1999), 251–260.

[21] J. B. Garnett, *Bounded Analytic Functions*, Academic Press, London, 1980.

[22] J. B. Garnett and P. W. Jones, *BMO from dyadic BMO*, Pacific J. Math., **99** (1982), no. 2, pp. 351–371.

[23] R. Hunt, B. Muckenhoupt, and R. Wheeden, *Weighted norm inequalities for the conjugate function and the Hilbert transform*, Trans. Amer. Math. Soc., **176** (1973), pp. 227–251.

[24] H. Helson and G. Szegö, *A problem in prediction theory*, Ann. Mat. Pura Appl., **51** (1960), pp. 107–138.

[25] P. W. Jones, *Square functions, Cauchy integrals, analytic capacity, and harmonic measure*, Harmonic analysis and partial differential equations (El Escorial, 1987), Lecture Notes in Math., vol. 1384, Springer-Verlag, Berlin–Heidelberg–New York, 1989, pp. 24–68.

[26] _____, *Rectifiable sets and the traveling salesman problem*, Invent. Math., **102** (1990), no. 1, pp. 1-15.

[27] _____, *The traveling salesman problem and harmonic analysis*, Conference on Mathematical Analysis (El Escorial, 1989), Publ. Mat., **35** (1991), no. 1, pp. 259–267.

[28] R. Kerman and E. Sawyer, *On weighted norm inequalities for positive linear operators*, Proc. Amer. Math. Soc., **105** (1989), no. 3, pp. 589–593.

[29] P. Koosis, *Moyennes quadratiques pondérées de fonctions périodiques et de leur conjuguées harmoniques*, C. R. Acad. Sci. Paris Sér. A, **291** (1980), no. 4, pp. 255–257.

[30] _____, *Introduction to H^p spaces*. London Math. Soc. Lecture Note Series, 40, Cambridge Univ. Press, 1980.

[31] S. Kupin, *Linear resolvent growth test for similarity of a weak contraction to a normal operator*, Ark. Mat., **39**, (2001), no. 1, pp. 95–119.

[32] P. Mattila, *Rectifiability, analytic capacity, and singular integrals*, Proc. of the ICM, v. II, (1998), 657–664.

[33] _____, *Search for geometric criteria for removable sets of bounded analytic functions*, Preprint, Jyväskyla Univ., 2003, pp. 1–18.

[34] P. Mattila, M. Melnikov, and J. Verdera, *The Cauchy integral, analytic capacity, and uniform rectifiability*, Ann. of Math. **144** (1996), 127–136.

[35] P. Mattila and P. Paramonov, *On geometric properties of harmonic Lip_1-capacity*, Pacific J. Math., v. 171, No. 2 (1995), 469–491.

[36] J. Mateu and X. Tolsa, *Riesz transforms and harmonic Lip_1-capacity in Cantor sets*, Preprint CRM.

[37] M. Melnikov, *Analytic capacity: discrete approach and curvature of measure*, Sbornik Math., v. 186, (1995), 827–846.

[38] T. Murai, *The Real Variable Methods for the Cauchy Transform, and Analytic Capacity*, Lecture Notes Math., vol. 1307, Springer-Verlag, Berlin–Heidelberg–New York, 1988.

[39] M. Melnikov and J. Verdera, *A geometric proof of the L^2 boundedness of the Cauchy integral on Lipschitz graphs*, IMRN (Internat. Math. Res. Notices) 1995, 325–331.

[40] N. K. Nikolski and S. Treil, *Linear resolvent growth of rank one perturbation of the unitary operator does not imply its similarity to a normal operator*, J. Analyse Math., **87**, (2002), pp. 415–431.

[41] F. Nazarov, *A counterexample to a problem of D. Sarason*, Preprint, Michigan State Univ., 1998, pp. 1–10.

[42] F. Nazarov and A. Volberg, *The Bellman function, the two-weight Hilbert transform, and the embeddings of the model space K_θ*, J. Analyse Math., **87**, (2002), pp. 385–414.

[43] F. Nazarov, S. Treil, and A. Volberg, *Cauchy Integral and Calderón-Zygmund operators on nonhomogeneous spaces*, International Math. Research Notices, **1997**, No 15, 103–726.

[44] _____, *Weak type estimates and Cotlar inequalities for Calderón-Zygmund operators on nonhomogeneous spaces*, International Math. Research Notices, **1998**, No 9, p. 463–487.

[45] _____, *Accretive system Tb theorem of M. Christ for nonhomogeneous spaces*, Duke Math. J., **113**, (2002), pp. 259–312

[46] _____, *Nonhomogeneous Tb theorem which proves Vitushkin's conjecture*, Preprint No. 519, CRM, Barcelona, 2002, 1–84.

[47] _____, *Tb theorems on nonhomogeneous spaces*, to appear in Acta Math.

[48] _____, *Two weighted T1 theorems*, Preprint, 2003, pp. 1–20.

[49] _____, *A criterion of two weighted boundedness of a Martingale Transform*, Preprint, 2003, pp. 1–10.

[50] _____, *The Bellman functions and two-weight inequalities for Haar multipliers*, J. Amer. Math. Soc., **12**, (1999), no. 4, 909–928.

[51] _____, *Counterexample to infinite dimensional Carleson embedding theorem*, C.R. Ac. Sci. Paris Sér. I Math., t. **325**, (1997), no. 4, 383–388.

[52] H. Pajot, *Analytic Capacity, Rectifiability, Menger's curvature and the Cauchy Integral*, Springer-Verlag, Berlin–Heidelberg–New York, 2002.

[53] V.V. Peller and S. V. Kruschev, *Hankel operators, best approximations and stationary Gaussian processes*, Uspechi Mat. Nauk, **37** (1982), no. 1, pp. 53–124.

[54] S. Petermichl, *Dyadic shifts and a logarithmic estimate for Hankel operators with matrix symbol*, C. R. Acad. Sci. Paris, Sér. I Math., **330**, (2000), no. 6, pp. 455–460.

[55] S. Petermichl, *Dyadic shifts and a logarithmic estimate for Hankel operator with matrix symbol*, Comptes Rendus Acad. Sci. Paris, t. 330, (2000), no. 1, pp. 455–460.

[56] S. Petermichl and A. Volberg, *Heating of the Ahlfors-Beurling operator: weakly quasiregular maps on the plane are quasiregular*, Duke Math. J., **112** (2002), no. 2, pp. 281–305.

[57] L. Prat, *Potential theory of signed Riesz kernels: capacity and Hausdorff measure*, Preprint, Univ. de Barcelona, pp. 1–48, 2002.

[58] E. Sawyer, *A characterization of a two-weight norm inequality for maximal operators*, Studia Math., **75** (1982), no. 1, pp. 1–11.

[59] _____, *Two–weight norm inequalities for certain maximal and integral operators*, Lecture Notes Math., vol. 908, Springer-Verlag, Berlin–Heidelberg–New York, 1982, pp. 102–127.

[60] X. Tolsa, *Cotlar's inequality and the existence of principal values for the Cauchy integral without the doubling condition*, J. Reine Angew. Math., **502** (1998), pp. 199–235.

[61] _____, *L^2-boundedness of the Cauchy integral operator for continuous measures*, Duke Math. J., **98** (1999), no. 2, 269–304.

[62] _____, *Cotlar's inequality and the existence of principal values for the Cauchy integral without doubling condition*, J. Reine Angew. Math. 502 (1998), 199–235.

[63] _____, *Curvature of measures, Cauchy singular integral, and analytic capacity*, Thesis, Dept. Math. Univ. Auton. de Barcelona, 1998.

[64] _____, *Painlevé's problem and the semiadditivity of analytic capacity*, Acta Math., **190** (2003), no. 1, 105–149.

[65] _____, *On the analytic capacity γ_+*, Indiana Univ. Math. J., **51** (2002), no. 2, pp. 317–343.

[66] S. Treil and A. Volberg, *Wavelets and the angle between Past and Future*, J. of Funct. Analysis, **143**, (1997), no. 2, pp. 269–308.

[67] _____, *Completely regular multivariate stationary processes and the Muckenhoupt condition*, Pacific J. Math., **190**, (1999), no. 2, 361–382.

[68] J. Verdera, *L^2 boundedness of the Cauchy integral and Menger curvature*, Cont. Math., vol. 277, Amer. Math. Soc., Providence, RI, 2001, pp. 139–158.

[69] A. G. Vitushkin, *The analytic capacity of sets in problems in approximation theory*, Uspechi Mat. Nauk, **22** (1967), pp. 141–199; English transl., Russian Math. Surveys, **22** (1967), pp. 139–200.

[70] A. Volberg, *Matrix A_p weights via S-functions*, J. Amer. Math. Soc., **10**, (1997), no. 2, pp. 445–466.

[71] J. Wermer, *Commuting spectral measures on Hilbert space*, Pacific J. Math., **4**, (1954), pp. 355–361.

Titles in This Series

TITLES IN THIS SERIES

For a complete list of titles in this series, visit the
AMS Bookstore at **www.ams.org/bookstore/**.